Around the World in 80 Years

Around the World in 80 Years

A lifetime of extraordinary people and adventures

PAT PETERSON

Storytellers Publishing
Colorado, USA

Storytellers Publishing
An imprint of Journey Institute Press,
a division of 50 in 52 Journey, Inc.
journeyinstitutepress.org

Copyright © 2024 Patricia Peterson
All rights reserved.

Journey Institute Press supports copyright. Copyright allows artistic creativity, encourages diverse voices, and promotes free speech. Thank you for purchasing an authorized edition of this work and for complying with copyright laws by not reproducing, scanning, or distributing any part of this work in any form without permission.

Library of Congress Control Number: 2024942237
Names: Peterson, Patricia
Title: AROUND THE WORLD IN 80 YEARS
Description: Colorado: Storytellers Publishing, 2024

Identifiers: ISBN 978-1-964754-03-1 (hardcover)
978-1-964754-04-8 (paperback)
978-1-964754-05-5 (ebook/kindle)
Subjects: BISAC:
TRAVEL / Special Interest / Adventure
BIOGRAPHY & AUTOBIOGRAPHY / Women
BIOGRAPHY & AUTOBIOGRAPHY / Adventurers & Explorers

First Edition
Printed in the United States of America

1 10 12 27 33 35 41 59 82 91

This book was typeset in Garamond / Fibon Neue

Cover Design by WiggleB Studios
Editing by: Jessica Medberry - InkWhale Editorial LLC

Photo Credit: All photographs used in this book are credited to the author and/or her family, and used with permission with the exception of Figure 1 on page xiv which is in the public domain.

CONTENTS

Preface   xi

Early Experiences Whet My Appetite for Adventure   15
    A Family Trip through the West   16
    Longs Peak—When Things Go Terribly Wrong   22
    The Maroon Bells—Mountains, Music, Marriage and More   26
    The Ultimate Waltz   29
    Who Can Tell the Best Fish Tale?   31
    Fishing in the Yukon River in Alaska   33

European Escapades   37
    Inspired by Switzerland   38
    Discovering Options   42
    Allure of the Open Road—An Introduction to Auto-Stop   44
    More Alluring Auto-Stops   47
    On the Road to Avignon: An Incredible Twist   49
    Auto-Stop in Brittany   51
    Boat Hitchhiking, or Bateau-Stop   53
    The Way to Luxembourg   55

A Greek Odyssey   57
    Prologue   58
    En Route   62
    Discovering Greece   64
    Return to Greece   67
    Crete   71

"O Give Me a Home . . ."   75
    Prairie Montana   76
    Jack Horner   79
    It Wasn't All Dinosaurs . . .   83
    Making the Most of Summers   87
    How about a Camping Trip in the Soviet Union?   89
    One Summer in Israel: "L'Chaim—To Life!"   96
    The Sinai Desert   97
    Jerusalem   101
    Family Adventures   103
    Slip Sliding Away   106

How I Moved beyond a "Whites Only" World   113
    Exposure in School   115
    A Trip through the Deep South   116
    Native American Adventures   118
    Florence Means   120
    Chikako Ando   121
    Summer Exposure in NYC   122
    Meeting Martin Luther King Jr.   124
    University Experiences   125
    Prejudice in Montana   127
    Buffalo Soldiers and Beyond   128
    A Reservation Encounter   130
    Hutterites   132
    Postscript   133

No Home Base—Nothing Permanent but Taxes   135
    Living in Divided Germany   136
    1989 and a Reunited Germany   140
    Mother Teresa   145
    Two Months in Bratislava   147
    The Gospel according to Mary   149

In Conclusion	151
Poland Is Open	152
Father Maximillian Kolbe	153
Grand Glimpses of Portugal	155
Zanzibar	157
Sister Mary: Witness and a Catalyst for Change	160

The World Is 70 Percent Water — 165

An Astronomical Event—January 2007	166
Cruise with Kir	167
A Rowboat Rescue	168
Church in the South Seas	169
Biking in Bali	170
Sailing in the Canary Islands	171
Scuba Diving around the World	173
Scuba Challenges	177
Here's to Small-Boat Adventure	180
One Week in a Canal Boat	181
Boating in the Pacific Northwest	182

Synchronicity Blows My Mind! — 187

The Telephone Call	188
Snakes	189
Frau Hanna Jolly	190
Paris	191
Animal Tales	192
Once upon a Hike	193
An Incongruity	194
What's Going On?	195

Retirement—Not Really! — 197

Haiti as I Lived It	198
First Impressions of Haiti	200
The Mission	202
Highlights and Disasters	204

Cuba Inside and Out	210
Japan in Person	213
The Episcopal Bishop of Okinawa	215
Rescue in the Grand Canyon	217
Mount Rainier	220
Around the World of Biking	223
Walking around the World	228
Nepal	230
Mount Everest	235
Walking and Volunteering in Colorado	237
Remembering Verlie	242

From Southern Hemisphere to Svalbard and Back — 245

Mishaps Make the Machu Picchu Adventure	246
Pat in Patagonia	251
Patagonia in Chile	252
Patagonia in Argentina	255
Two Months in South Korea	262
Bosung Girls' Middle School	265
Jirisan—Sacred Highest Summit	269
A Weekend at Jeju Island	272
Final Impressions of Korea	274
Kilimanjaro: On the Roof of Africa	276
Eighty Degrees North Latitude	282
2023—The Grand Finale	284
The Road Not Taken	291

For those often unrecognized people
who engage in acts of kindness inspiring
our voyage through life around the world.

Preface

I am an octogenarian and an active widow. I enjoy, among other things, volunteering, hiking, classical music, travel, and meeting people. I have been encouraged to write about the adventures and relationships that have enriched my life over the years and around the world. Some stories are begging to be told. Some people are too remarkable to forget. Some incidents embody the very gist of life. Some of them almost cut mine short! I am grateful to take the reader with me as I relive these events.

This is the unfolding story of my life. My family settled in Boulder, Colorado, which I have called home for much of my life. In my late twenties, Hans Peterson (Hajo for short) and I married in Berlin, Germany, his native land. Eventually, we made our mutual home in rural Havre, Montana, for twenty years. In all, we lived over fifty adventure-filled years together! More recently I have been calling Boulder my home again. In between these periods in Colorado and Montana, I have travelled, lived, and worked around the world for about twenty-five years.

I am the oldest of four children. Sue is four years younger than I, and she became a teacher and later a college administrator. Mike is four years younger than Sue, and after years as a cross-country and US Army biathlon skier, he became an architect in Seattle. Our youngest sister, Jane, was a teacher in Alaska until her untimely death from cancer in her early forties.

I was fortunate to grow up with parents who loved the outdoors, camping, and hiking; who stressed getting a good education; and who taught by example the importance of reaching out to people of different backgrounds. My father, Stephen Romine, was a university professor and dean. We four children called him Pop. Remarkable for the time, he showed me that a girl could do anything that a boy could. For example, he took me rope climbing on the Boulder Flatirons, the signature rock formations overlooking the community, in my early teens. My mother, Marguerite, taught us an appreciation for music in general, and opera in particular, as we listened together to the Metropolitan Opera Radio on many Saturday mornings. She and Pop expanded our vision. Our parents invited two exchange students to live with us, each for a year, from Japan and Spain.

Our grandmother lived with us part of the time. Known as Gran, she was our only living grandparent. We loved her tales of Greyhound bus trips around the country. Her parents had been German immigrants to the US, and she could still speak German, which she had spoken as a child. When I was about ten years old, I asked her to teach me some. She said, "Okay. *Ich fahre nach Deutschland.*" I said, "Gran, why are you teaching me 'I am going to Germany?' I'll never go to Germany." She shrugged and replied that it had just come to her, and thus I learned the phrase. Was she prescient, perhaps? Fifteen years later, I did travel to Germany and married Hajo in Berlin. Later, I taught English as a foreign language in Berlin, Dresden, Halle, Frankfurt, and other German cities.

Hajo and I had two sons, Eric and Mark. I call Eric my ocean son because he loves the sea. We used to go scuba diving together, and we continue to enjoy spending time on his live-aboard motorboat and touring around the Salish Sea in the Pacific Northwest. He worked in the tech world until he became a winemaker for Patterson Cellars. He is married to Lanell, who works in cancer research. They have two grown children—my grandchildren, Brendan and Kayleigh. Mark is my mountain son, and we often hike together. We have climbed Mount Rainier and trekked in Nepal. He has his own public relations firm and a variety of clients. Our family gets together several times a year for grand reunions!

I have titled this book "Around the World in 80 Years." "Around" means both travelling around the globe and living around the United States in between trips abroad. I have chosen these stories for their unique character, their adventure, and the people involved, and because I believe they need to be told. Not everything I write about is hunky-dory. Difficult realities engulf our world, and these are often part of my stories.

As I am often reminded, I won't live forever. If I don't record these stories, they will be lost, and the world that much poorer for the loss. I am glad to be old; the other option is pending and far less appealing. I love life and agree with the sign I saw in Blanding, Utah: "I thought that getting old would take longer." I've also learned that old age is a matter of perspective. I thought I was old until I started driving a group of seniors to weekly university concerts. They are ninety-eight, ninety-six, ninety-four, and ninety. They call me "youngster." I know I have slowed down from my road-running and racing days due to heart problems and other issues. People sympathize with me. Don't they know there is life after running! My only race now is against time, to complete this volume before I can't!

Older and wiser? There is a myth that elders automatically have insights into life simply by being old. Maybe I am still too young, but I don't know the answers to the overwhelming challenges of life: injustice, violence, prejudice, global warming, selfishness, egotism, homelessness, hunger, illness, and apathy, to mention just a few. I continue to seek meaning and perspective through Christian, Jewish, Islamic, Buddhist, and other spiritual expressions and practices. I like to follow the example of the pilgrimage, perhaps a modern *Canterbury Tales*. As we move along the paths of life, we share perspectives with others, show kindness, express gratitude, and meet needs when possible and appropriate. I don't feel called to convert others to my brand of beliefs. However, I do hope that together—through a spirit of openness, mutual caring, and respect; sharing personal gifts; the taking of responsibility for planet earth; and a sense of humor—we can all make the world a better place. These are my stories. Happy reading!

Figure 1. Family carrying the forest-cut Christmas tree circa 1953

Figure 2. Gran and granddaughter Pat all dressed up (1945)

Figure 3. Flatirons: Backdrop to Boulder

Early Experiences Whet My Appetite for Adventure

A Family Trip through the West

The summer of 1951, our family set off from Boulder to camp our way through the American West. We planned to visit National Parks and places in between. This was quite an expedition, with two parents, one grandmother, and four children ages thirteen, nine, five, and three. We were all crammed into our black-and-chartreuse Chevy station wagon, which pulled a small one-wheeled trailer full of our camping gear. I anticipated a fun time, but it was so much more! Looking back now, seventy-plus years later, it's a wonder that our mother managed all the chores and children while Pop drove the car for four weeks, setting up and tearing down the tent every few days. Mother read to us in the back of the car, which made the miles of travelling tolerable, and Gran helped along the way. This story is not a travelogue, but a telling of the salient experiences I enjoy remembering to this day.

Our first stop was Salt Lake City and the Tabernacle Square, where I learned the stories of Joseph Smith and the Mormon pioneers. I collected some of the free pamphlets and wanted to read about this religion, which was new and tempting to me. I didn't want my family to know about my interest, so I asked Pop if I could sleep out of the family tent, under a tarp in the woods nearby. He fixed my sleeping bag under a canvas cover. At bedtime, I crawled in with my secret Latter-Day Saints literature and a flashlight. There, under cover, I began to read about the Book of Mormon and Joseph Smith. I had not read for very long when I thought I felt something near my head outside the tarp. No, maybe it was the wind? I kept reading. I was sure there was something by my head. I carefully shined my flashlight out of the sleeping bag. I was face to face with a porcupine. When he saw the light, his quills rose up around his face, a few inches from my own! I scooted down in my bag and held my breath. When I finally looked out again, I could see his tracks in the dust heading away into the dark. I called loudly, "Pop, Pop, Help!" He came to my rescue, and I told him about my encounter with the porcupine. I rejoined the family in the tent. Thus, I was saved, not by Joseph Smith, but from a porcupine!

The next day, we headed for Yellowstone National Park. I remember long lines of cars stopped as people fed bears out of their car windows despite park regulations prohibiting it. Park visitors were told they could see the bears eating at the park dump every evening. The whole thing seemed unnatural to us. We camped at Madison Junction, where the bison meandered freely around the tents. Somehow, this was not considered dangerous. Mother washed clothes and shampooed our hair in warm springs along the Firehole River. That seemed natural in those days.

We waited expectantly on the small boardwalk by Old Faithful. It never failed to erupt and thrill us with its tall jets of water shooting straight up, the spray droplets glistening in the sun. A ranger bulletin board announced the expected eruption of many other geysers, and we hiked or drove to see them, discussing which was the tallest, or most beautiful, along the way. We loved strolling around the hot springs, especially the Morning Glory Pool with the clear blue color and shape of its namesake. Pop tried to catch a fish in Yellowstone Lake to cook in a nearby hot pool, but the fish were not biting that day!

Yellowstone Canyon and the Lower Falls were a surprise for me. I had never heard about them. Imagine walking up to a lookout point and seeing, for the first time, a gigantic three-hundred-foot waterfall dropping into a deep canyon with pastel pink and golden walls. The roar of falling water and the rush of a green river below filled the air. I stood there with my family, totally astonished! We stopped at all the special canyon-viewing sites. Near the base of the Lower

Falls, the water crashed with earth-shaking power, creating with the sunlight a bright rainbow in its spray. I realized that this golden canyon had made Yellowstone the name of both the river and the national park.

After Yellowstone, we drove south to Jenny Lake and the Tetons. In those days, a family could make an early start and count on finding a camp spot in one of the several campgrounds. No reserving months ahead, as campers must do today with the population explosion. We loved Jenny Lake and the rugged peaks rising from its clear waters. Pop and I decided he and I would go mountain climbing while the rest of the family enjoyed the lake, ranger hikes, and evening campfire programs.

Our first climb was up Mount Teewinot, more than five thousand vertical feet up from the valley floor. It was a steep, hard scramble over giant boulders and up cliffs on its eastern face. We were in good shape and enjoyed the climb in the warm, sunny weather. But as we climbed the last pitch to the summit, we were suddenly confronted by the dark menacing clouds of a thunderstorm moving in our direction. They were totally unexpected and alarming! Before we could set up a rappel to help us get off the mountain quickly, we heard the clicking of electricity and saw the light-blue sparks of Saint Elmo's fire skipping over the summit rocks. Bright bolts of lightning and terrific crashes of thunder hit simultaneously, dangerously close. I could tell Pop was frightened, and I was terrified. We managed to rappel down the rope and away from the worst of the electric storm. Conditions were perfect for us to be electrocuted, but somehow we survived! We stumbled down the mountain and eventually to the tree line, where we felt safer as the rain began to fall. We were drenched, but happy to be alive. It was here that I learned not to take thunderstorms lightly!

The summit of the Grand Teton was the goal of our final ascent. Many climbers hire a guide for this mountain, but Pop had a book that detailed the particulars of the route he had chosen for us. We set out in the afternoon to hike around the "Grand" to the west side, where the actual climb began. We bivouacked that night in the presence of stark, dramatic peaks that rose above us like sentinels guarding our simple camp. I felt so small as I fell asleep in this world of pinnacles and stars.

We set off the next morning full of anticipation. We roped up because it was a very exposed climb, but with solid hand- and footholds. I loved the sensation of gripping such trustworthy rock as I climbed. I could look through my legs and see lots of empty space below me. I decided then that it was better to look up! It was exhilarating to climb such a spectacular mountain. We shared the summit with several other climbers who had ascended by another route. After greetings and mutual congratulations, we sat there in sublime silence, awed by the setting, with only the sky above us. Then we broke into our packs and shared trail snacks washed down with canteen water. We all signed our names in the summit register we found rolled up inside an iron pipe wedged into some rocks. Pop said I was probably one of the youngest climbers and maybe the only girl to summit the Grand at that point. I felt a certain pride, but mostly I was overwhelmed by the euphoria of climbing such a magnificent mountain.

From Wyoming, we drove to the Pacific Northwest. I had never seen a real ocean before. I was mesmerized by the roar of the waves crashing on the rocky headlands and the breakers rolling up the broad stretches of sandy beach. While camped one night, we fell asleep listening to the drone of a foghorn through the cloudy, rainy darkness.

We continued south and inland to Yosemite National Park, where we camped on the valley floor. Great waterfalls dropped from impossibly high cliffs around us, but what I most remember is the Fire Fall. The rangers built a huge bonfire on the brink of a cliff at dusk. When the fire was at its roaring red peak, it was gradually pushed over the precipice, creating a long, blazing cascade of embers dropping down through dark space: the Fire Fall! Though it was cheered by hundreds of viewers, it was discontinued, as it was not a natural wonder of the park.

Several members of our family climbed to the top of Half Dome by clinging to the cables along the barren-rock route. Though I had already climbed many mountains and been more exposed, for some reason I had terrible vertigo up there. I refused to sit on the edge of a rock jutting out over the canyon and dangle my legs into space, as some daredevils were doing!

From Yosemite, we travelled to visit the giant trees of Sequoia and Kings Canyon. Our family all took hold of hands, trying to encircle one of the huge trunks, but we reached less than halfway around it. On the most memorable day, we were hiking among the giants when we heard a deafening "Kaboom!" Somewhere not far away, one of the ancient trees had fallen! Pop led us in the direction of the terrible sound, where we found the splintered trunk of the shattered giant. Standing beside the remains, I could not even see over the side of it. How sad yet moving to have witnessed the demise of such an ancient monarch of the world. Seventy years later, I still feel humbled remembering the death of this two-thousand-year-old tree whose time had come to a natural end.

Mesa Verde National Park, with its cliff dwellings and other structures, was the concluding experience of our summer camping tour. I was amazed by the master builders of the towers and kivas, all constructed by hand. I caught the archaeology bug as I listened to the rangers explain the particulars of the Indigenous people's lives so many centuries earlier. It was all so fascinating, and it has remained so to this day in my exploration of numerous ancient sites throughout the Southwest and the world. I realize now that this family trip really inspired me to travel and see firsthand the natural and cultural wonders of the world in all their diversity.

At the same time, I have realized I want to do my part to preserve these valuable resources by participating in multiple volunteer projects: building and restoring trails, eradicating invasive species, collecting and destroying pollutants and plastic, leading student field trips, and participating in cultural and scientific excavations around the world. Thanks for your influence, Mother and Pop! And thanks to the many ecological organizations working to recognize and preserve the natural and cultural treasures of planet Earth. They deserve our support!

Longs Peak—When Things Go Terribly Wrong

Longs Peak dominates the other mountains in northern Colorado. Its barren east face rises a threatening three thousand feet above Chasm Lake, the granite sheared off by centuries of eroding forces—ice, wind, and water brutally scouring away ledges or any place a seed might find to root and grow. Maybe this stark demeanor is what draws climbers to match wits and brawn against its utterly bleak challenges. Whatever the case, Pop felt the lure and drew me, his fifteen-year-old daughter, to climb Notch Couloir with him. This route rises south of the east face, several thousand feet of dramatically stacked boulders with adequate foot- and handholds and belay places, leading right to the summit... that is, if the weather is fine. But this mountain is notorious for catching storms and even brash blizzards.

Pop and I started up the well-groomed trail leading to the mountain itself in the early morning darkness of a September Sunday. Our headlamps lit the five-mile trail to Chasm Lake, which we skirted to the south. We found the suggested route up a steep couloir as the sun slowly rose in an overcast sky. We clambered up a steep moraine and then crossed over the sliver of a snowfield, or was it a glacial remnant? It was hard ice. Pop chopped steps across it with an ice ax, and we roped up across it for safety. We then made our way to a long, broad ledge cutting across the base of the east face, appropriately called Broadway! By this time, the overcast weather had turned to roiling clouds. A few wind-driven icy flakes brushed past us. Suddenly, gusts of wind began blasting us with masses of wet snow. A blizzard on Broadway! We huddled together behind a boulder, waiting out this unexpected challenge from

Longs Peak—When Things Go Terribly Wrong

the mountain. An hour later, the storm blew itself out, leaving two inches of icy snow on every rock, crevasse, and surface.

Pop said, "We shouldn't go back the way we came. That little glacier is too steep, and it will avalanche with us if we try to cross it. The only way to go is up."

"But," I protested, "there is ice on everything."

"We'll have to be careful, but we can do it," Pop said with more confidence than I felt as my boots slipped on the ground when I tried to stand. "You will take the lead," he continued. "I'll show you where to go. I will belay you. If you fall, I will be able to stop you. That will be better than your trying to stop me if I fall. We'll go on short rope lengths."

I fixed a rope harness around my waist and thighs. We made our own harnesses in those days. Pop checked the knots to be sure they would hold. The belay rope was secure, and Pop was positioned in such a way that he would not be pulled off the mountain if I fell.

I began climbing. I knew enough to keep three secure holds before I moved to a fourth. Two footholds and one handhold, two handholds and one foothold, before I moved up or brushed snow off a potential step up. I wanted to wear my gloves, but I needed bare hands to grip the handholds. I kept them to wear when I wasn't climbing.

The first leads were easy, with big, deep footholds I could kick clear to take me up the route as I balanced myself with clear handholds. I moved steadily up and could look down about thirty feet to the right to see Pop paying out the rope as I climbed.

He called up, "Do you see a good belay place for you to bring me up to you?"

I answered, "Not here, but maybe up above the next stretch. I'll try to find one."

I attempted to kick clear another foothold, but it was covered in very hard ice. The right foothold sloped out and was slick. But my handholds were good. I was maneuvering to find a better foothold when suddenly both feet simultaneously slid out from under me. Then the rock I was gripping with my left hand pulled off the mountain and I was left hanging from my cold, numb right hand alone. I fought to find purchase. My right hand slid off its hold. I was falling, uncontrollably falling!

I dug my fingers into the rock, trying to slow my chute. Every inch of my body seemed to be screaming, "Stop falling. Stop falling." I felt a self-preservation instinct like never before. My shoulder slammed into a rock, and I half turned as I slid past Pop. It seemed like the color drained from his face as I dropped past him. I felt like I was tearing off my fingerprints on the boulders as I pressed my fingers into anything to slow down. I couldn't find any purchase with my clumsy boots. I knew that for the rope to stop me, I had to fall as far past Pop as I had been above him when my fall started. Suddenly there was an abrupt jerk on the belay rope harness around my waist as I came to a violent stop.

"Are you okay?" Pop called down.

"I think so," I called back up. My legs were shaking with adrenaline. But my feet had landed on a solid flat space.

"Stay there. Don't move. I am coming down."

I hated that part, seeing Pop working his way down to me with no protection if he fell. My shoulder was hurting, but not seriously injured. I felt guilty for having fallen, but I'd done my best. I couldn't help it. I apologized to him. He said, matter-of-fact, that we still needed to get up and off the mountain. More snow could blow in.

He took the lead on the rest of the climb. He found me secure places to belay him and sometimes tied me to the cliff so the rope would stop him even if I couldn't alone. I continued shaking most of the time, whether from nerves or cold, I don't remember. I just had to tough it out and do my best to belay Pop. Fortunately, he did not slip or fall as he cautiously climbed above me. I was able to follow up to him without falling again.

We made it to the summit but didn't stop to celebrate. We headed over to the north side of the mountain where the cable route and painted bulls' eyes still existed, now long since removed. We easily maneuvered down and were soon back on the boulder field working our way to the trail, an easy if long path back to our car. I was exhausted, but just kept plodding along behind Pop. He had taught me to be tough, and that kept me going. I wondered if my mother would be worried when we were so late.

We reached the car in the dark. It had been a very long day. I slept as Pop drove home. When we reached the house, I said a hasty hello to Mother and went right to bed. I had school the next day.

Monday morning, I showered and looked at my sore hands and the scratches and bruises on my legs. I had been lucky. I went off to school like nothing had happened, and nothing was said. I figured Pop had told Mother, but neither of them mentioned it. I may have fallen, but my Pop's belay had saved me, so everything was fine. No harm done.

Only years later did Mother learn of my fall. She was furious that Pop had said nothing. What if he had fallen instead of me and left her a widow with four children? Pop rationalized some things, and I just went along for the adventure. Soon my younger brother was old enough to be his climbing companion, and I gave up mountaineering.

The Maroon Bells—Mountains, Music, Marriage and More

I am not sure what iconic means in this case, but I have heard that the Maroon Bells are some of the most iconic mountains in Colorado. I do know that they are among my most-loved peaks—for their maroon color; for their reflection in Maroon Lake; for their inspiring, splendid photographs; and for the memories of so many events in my life that they evoke. The Bells, as they are often called, symbolize the spirit of adventure that has motivated much of my life since childhood.

As a family, we often crammed into our well-used black-and-chartreuse Chevy station wagon and drove from Boulder over multiple passes to Aspen. From there it was ten more miles of dirt road to Maroon Lake campground. On one trip, we chose an unoccupied site near the stream outlet to the lake, where Pop pitched our square green canvas tent. Then he vanished to go fly-fishing in the lake, leaving Mother and us four kids to fend for ourselves. We kids collected small logs from the beach and wrapped them in ropes to make a rough but navigable raft, which my two sisters and brother paddled into the lake while I stood on shore catching the moment in a Brownie box camera. Today, we still look at the black-and-white photo with intense nostalgia.

It was some twenty years later that Pop would be fly-fishing in this same spot and look across the eastern end of the lake to see another fly fisherman standing in waders and on crutches, casting in his direction. Pop was puzzled. The angler looked familiar. Who could it be? Then he knew: Itzhak Perlman, the world-renowned violinist whom he would hear that evening at the Aspen Music Festival, was fishing with him in Maroon

Lake. Maestro Perlman, with casting rod and line, exuded the same precision and joy he expressed with violin and bow in his concerts. He waved at Pop, and Pop waved back . . . unity of spirit at Maroon Lake.

Some years later yet, the whole family gathered again, this time for Jane's daughter Willow's wedding. As we waited for the ceremony to begin, we each had our own memories of these mountains. I relived climbing the fourteen-thousand-foot peaks as a teenager. Pop and I had maneuvered up nearly vertical pitches composed of decaying maroon cliffs and perilously balanced boulders. There was no trail. The process was treacherous. At one point, I remember reaching up to take hold of a boulder the size of a grand piano. It was so loose and the gradient so steep that it slid past us and tumbled down the precipice, coming to rest far below us. Fortunately, we were the only climbers on the mountain that day!

My attention returned to the wedding. My niece was radiant in a simple white gown as she was escorted forward by her father and stepmother. The ceremony and reception, surrounded by nature, were perfect. We missed Jane, her mother, but somehow we sensed she was with us in spirit.

I continue to enjoy returning to the Maroon Bells. A few years ago, I volunteered with a seed-gathering group dedicated to preserving native flower species. It was late summer as we collected seeds and labelled lupine, marsh marigolds, mule's ears, columbine, paintbrush, and dozens of other seeds. We noted exactly where we had found them and the growing conditions of each plant so the seeds could be planted elsewhere and have a chance of flourishing.

Nothing tarnishes my cherished memories of the Bells. Watching a beaver add branches to his home at the end of the lake. Seeing my mother sitting on a large smooth rock in a flower-filled meadow, smiling while gazing up at the Maroon Bells. Pop making his final tent camping trip there in his nineties. He wanted to camp on the ground, to feel the earth beneath his body. The next morning, it took his son to help him to his feet as he crawled out of the low door with a big grin on his face . . . his camping wish fulfilled.

Perhaps the growing numbers of grandchildren will also travel to the Maroon Bells to create their own adventures and stories in the family tradition.

The Ultimate Waltz

My early years were about more than camping and climbing. When I was a high school student, I also nurtured dreams of becoming a ballerina. My dreams became a reality for one short hour on a spring Saturday in 1954.

My ballet teacher had rehearsed with my dance partner Dan and I, choreographing a grand waltz to a Strauss melody. We practiced in the small room of our dance studio until Madame declared we were ready.

We piled into a van with our costumes, a large Victrola, and a record of our Strauss waltz. We drove to the magnificent Red Rocks Amphitheater near Morrison, Colorado. Great red sandstone formations create a dramatic setting for both performers and audience. In those days, everyone had access to the whole area. We lugged the antique Victrola to the side of the stage. There were no electrical connections at that time, nor were they needed. We carefully smoothed our costumes and put on our ballet shoes. Madame cranked up the Victrola and put the needle on the record. *Sounds of Strauss* reverberated around us: Vienna Woods, transported to Colorado mountains.

The two of us were poised for the first downbeat, and then we began our much-rehearsed waltz, circling the stage. The grand red cliffs rose dramatically above us to the left and right. Ta dum, ta dum . . . Dan took my right hand, twirling me around. My turquoise gown flowed rhythmically as we whirled together, dipping and springing under the open azure sky . . . the sun our only spotlight, a few surprised tourists our only audience. It all seemed so effortless, so perfect, as my partner lifted me and I was suspended in sheer joy. Then we were gliding across the floor in graceful synchronization. But the Victrola slowed, and then slowed a bit more, and stopped. The magic had come to an end. My partner bowed an adieu and I curtsied. Standing there without music and motion, we seemed so ordinary. But, for a few inspired minutes, we knew the exhilaration of dancing at Red Rocks, and our teacher was gratified to have given us the opportunity.

Who Can Tell the Best Fish Tale?

I grew up hearing tales from Pop and his fishermen friends . . . like the one about casting a fly over a lake and accidentally catching a swallow that was out eating insects one evening. They reeled in the bird and removed the fly from its beak, and it flew away. Or the time they simultaneously caught three fish on a line of lures while trolling on a boat. Or when they caught the same fish twice in a catch-and-release pond.

I must admit that I have ambivalent feelings about fishing. I love watching trout drift peacefully in quiet pools. On the other hand, I know the thrill of having a fish on my line and managing to land it! And I do love fish for dinner! Pop tried to teach me to fly fish in Boulder Creek before there was a kiddie pool in town. I cast my flies very well, right into the bushes. He untangled my line many times as I became more and more frustrated. Finally, when I was twelve, he said I was ready for a real backpacking fishing trip. He took me to Ypsilon Lake, tucked in the forest below Ypsilon Mountain in Rocky Mountain National Park. We set up camp near a stream flowing into the lake. Then, Pop stationed me on a large boulder by the lake without any bushes in sight. I cast a few times without a strike. But then I saw a trout rise and head for my fly floating on the surface. I was excited! The trout looked at my fly, turned away, and dived out of sight. Suddenly I felt a tug on my line. My pole bent down. I had a fish on!

I called to Pop, who ran over with his net, and together we landed my first lake trout. To my surprise, my fly was not caught in the fish's mouth at all. The fly was hooked in a nylon loop of a leader sticking out of the fish's mouth. I had lassoed my first fish. Later we discovered the original hook disintegrating in the stomach of my catch. I felt genuine sympathy for the poor fish.

Fishing in the Yukon River in Alaska

Many years later, in 1995, I was visiting Shirley Johnsrud, a close friend from my Havre days. She had moved with her husband Dave to Alaska because "Montana was not wild enough." Dave had built a cabin out on the Yukon River west of Galena, the town where they made their home. Shirley and I were going to rough it in the cabin on our own while Dave returned to carpentry jobs in town. The first morning, when Dave was still with us, I heard a strange noise at the cabin door. I said, "Dave, there's something at the door!"

Dave replied, "I don't hear anything. You must be spooked, being out here in the wilderness this way."

I waited a few minutes and timidly suggested, "I still hear something at the door."

Dave gave in: "Well, I'll check." He returned from the door swearing, "It's that damned porcupine! He ate my last cabin door. He won't eat this one!" With that, Dave clobbered the porcupine with a crowbar.

I was stunned, but said, "I've heard that porcupine meat can save your life if you are starving in the wilderness."

"Okay," Dave replied. "We'll have hindquarter of porcupine for lunch. Too bad we don't have a can of beans. Then we could have pork and beans!"

Later we roasted Porky. Dave sliced thin pieces of hindquarter, which we chewed and chewed and chewed. It tasted "like chicken" but was so tough a guy living off the land would have starved to death before he could swallow it!

Shirley and I would have to fish for our protein!

But fishing was not all that easy. We fished for a week without a single strike.

The last morning of our stay, I told Shirley I'd try fishing again. I climbed down the twenty steps of a wooden staircase anchored to the steep bank of the murky Yukon. From the small landing dock, I cast a spinner way out. Before I even began to reel it in, I had a tremendous strike! The battle began. Whatever was on the line was diving down and jerking back and forth from side to side. I gave the line some slack and saw the large silvery body rise to the surface. I knew I couldn't land it alone. It might even break the line. I finally found a way to brace the fishing pole between two tree trunks. I left it jammed there, climbed the stairs, banged on the cabin door, and shouted, "Shirley! Help! I have a fish on!"

She grabbed the net, and we raced down to where the fishing pole was still braced between the trees. Ten minutes later, we had landed a twenty-pound sheefish on a ten-pound test line. These delicious white fish migrate up the Yukon every summer. The migration was just beginning. We didn't have a stringer on which to attach the fish, so we substituted a logging chain that we found under the cabin.

Before Dave came to pick us up at day's end, we had caught two more sheefish. Two had broken our lines and gotten away. Fish fillets never tasted so good, and Dave and Shirley had fish for the freezer and for winter use.

Incidentally, a major flood on the Yukon some years later swept away the cabin. It was photographed still floating over fifty miles down the river, recognizable by the unique roof and chimney.

Very sadly, Shirley later drifted into Alzheimer's and had no memory of our fishing adventure, or of the trip we spent exploring Europe together, or of teaching and hiking for many years in Montana. What a terrible disease. It finally took her life in 2023. But how I treasure our many special times together. A true friend.

Figure 4. Two summits: Teewinot, left, and Grand Teton, center

Figure 5. My family skirting a California Redwood

Figure 6. Storm ravaged route up Long's Peak

Figure 7. Siblings rafting on Maroon Lake

Figure 8. Mother enjoying the Maroon Bells

Figure 9. Pop fly fishing

Figure 10. Last day's catch on the Yukon River

European Escapades

Inspired by Switzerland

When I was a young teenager in Boulder, I discovered a book tucked into one of our living-room shelves. It contained a photo of the Matterhorn. This magnificent, stark mountain captured my imagination. I loved hiking in the Colorado Rockies. Seeing the image of this Swiss peak inspired me to start saving money for a trip to Switzerland. This goal was fueled further by my favorite book, *Heidi* by Johanna Spyri. Like in this idyllic alpine story, I could imagine myself hiking through summer meadows and up mountains with Peter the goatherd.

I was a junior at the University of Colorado before I had saved up enough money to plan my trip. I had lived at home in Boulder to save money while attending the university. I heard that the Experiment in International Living provided partial scholarships for living with families abroad. I applied and was chosen for Switzerland with a $300 stipend. That, with my savings, $100 from my parents, and several gifts, gave me the $1,000 I needed for the adventure. It included round-trip boat travel to Europe and back on a refurbished Dutch troop ship, train travel through Holland and down the Rhine to Switzerland, three weeks living with a farming family near Lake Geneva, three weeks travelling and hiking through Switzerland, and one week in Paris. It seemed like a dream come true, and it was!

My Swiss family, the Lequints, lived in Givins, a village of three hundred near Nyon and Lake Geneva. They were a farming family living almost exclusively on what they raised themselves, with wheat fields, vineyards, hay pastures, milk cows, pigs, a vegetable garden, and fruit trees. My Swiss mother was from the German-speaking area. She had married my French-speaking Swiss father. They had two children: Jacqueline, my teenage sister, and Henri, her younger brother. The family hired four farmhands, three young people from the German part of Switzerland and one middle-aged woman from the Italian area, to work in the vineyard. Mealtime was a multilingual party as we all tried to communicate in four different languages. The family had a tractor but no car, so we biked, walked, or took the small local train to further destinations. Looking back, I am glad to have lived in such a self-contained family and society. I raked hay, pruned grapevines, and tended the garden with Jacqueline and the women. I felt like I was living in a traditional society that went back hundreds of years.

My Swiss father was required to go for military exercises every few weekends as part of his duty to preserve Swiss neutrality. His wife did not have the right to vote. She didn't mind because she said that she told her husband how to vote. They were proud to be Swiss, but equally proud to be from Canton Vaud, one of the Swiss states. We all sang the Canton Vaud song, but I can't remember learning a Swiss anthem.

Switzerland may seem isolated from the rest of the world, but it was there that I had a firsthand experience with someone who had survived the Nazi blockade of Leningrad during World War II. A friend of the Lequint family came for tea one afternoon. She was a retired teacher. She had worked for many years as a governess for the children of a family in Saint Petersburg (later Leningrad under the Communists). She was considered an essential part of the family in Russia, especially when the men were conscripted into the Soviet army. Then, in September 1941, the Nazis surrounded the city, cutting it off from the outside world. Hitler's goal originally was to take over Leningrad by destroying its industries and its symbolic importance as the cultural center of Russia. However, the planned quick capture of the city failed, and the Nazis resorted to war by starvation. In all, the blockade of Leningrad lasted for almost nine hundred days.

The first winter was the most brutal, with extreme cold and next to nothing to eat. Our tea party visitor in Switzerland told of the rationing of food, a kind of bread made with very little flour, fine sawdust, grass, and anything to give it substance if not nutrition. She explained, "Each person was allotted several slices of this bread per day and nothing else. The adults often shared these meager rations so the children could have more to eat. It was terrible, but we traded our house cats around the neighborhood so we could have cat stew from another apartment while they ate ours. We made up stories when our cat disappeared. We resorted to burning chairs, floorboards, even books to have a bit of warmth in one room where we huddled together. I continued to teach the children just to keep their minds on something besides the daily horror of our existence. Over one million people died of starvation in the siege of Leningrad. Fortunately, after the first winter, the family and I found places on a Soviet-organized convoy out of the city, so we survived. The family, who was like my own, travelled to family members elsewhere in Russia. As a Swiss citizen, I was considered neutral and made it back to Switzerland."

I was spellbound by her descriptions. I had studied Russian history at the university. For the first time, I was profoundly thankful I had studied French and could hear and understand the human stories I would otherwise have missed.

Our group of ten Americans with the Experiment joined our Swiss brothers and sisters for a three-week tour of Switzerland. We enjoyed the picturesque old cities of Bern, Lucerne, and others; the international centers of Geneva; and the historic tours of castles such as the Château de Chillon. But my favorite times were in the Alps, in high villages such as Arosa and especially Zinal in the canton of Valais. Here, we girls slept in a youth hostel hayloft. We ate raclette, an area specialty of melted and partially crispy cheese scraped over potatoes. We hiked trails, finding edelweiss growing in the meadows while listening to the cowbells echoing across the valleys. I even found Peter, a Swiss teenager who had grown up scaling mountains. He led me on a full-day climb up the three summits of Les Diablons, ropes and all. We had an idyllic time and even saw the Matterhorn far in the distance. My dreams of Switzerland had become a reality.

Our last week in Europe was spent in Paris, an extraordinary cultural change from the rural villages. We stayed on the Left Bank in a student hotel and walked all over the city, from Montmartre to the Eiffel Tower, to Notre Dame and through the Louvre. We were gorging on history and culture. I remember watching the Seine flow along beside the quays and thinking, "This whole European experience is so fulfilling, so wonderful. I must come back and live here or else never come back at all. This small taste of travel is too tantalizing to come for short appetite-whetting stays." Thus, I have spent much of my life travelling, living, working, volunteering, and making my way around the world.

Over the years, I returned to Switzerland many times. My husband and I hiked in the shadow of the Matterhorn with baby Eric in a backpack. Hiking alone in the fog on another trip, I suddenly came face to face with an ibex whose long, curved horns were intimidating! I backed up cautiously. I knew what to do if I met a bear, moose, or mountain lion, but an ibex? Fortunately, he was more interested in nibbling grass than me as I circled around him and up the slope. My last Swiss trip was to Zurich with friends from former East Germany whose son was studying there. The world was becoming ever more international!

Discovering Options

After Switzerland, it was two years before I had an opportunity to live and work abroad.

It was the autumn of 1961. I had just arrived in Fontainebleau, France, to begin my year as a Fulbright teaching fellow in the Lycée de Jeunes Filles, a French girls' high school. I was one of about twenty-five new university graduates who were each assigned to a different high school around France to teach spoken American English. We were to teach three days a week and attend classes at nearby universities the rest of the time.

The Sorbonne in Paris was the nearest university to Fontainebleau. I dutifully applied and was accepted. I underwent the required health examination. This entailed standing in lines to be weighed and measured and jabbed by sticking my arm through a hole in a wall to have blood drawn by an unseen technician. I received my university student card and stood in another line for entry to the university cafeteria. There was such a mob of students that I was jostled around and had the belt of my coat torn off and my watchband broken before I got my tray. I hurriedly ate lunch and headed for a lecture by a recommended professor. The room was jammed, and I found a place in the corridor, trying to listen through the open door. I could hear almost nothing and left after learning I could purchase written texts of the lectures I missed.

I returned to the quiet of provincial life in Fontainebleau. I was welcomed by my fellow teachers. I found my identity as Patricia—not Pat, which in French was *patte* (an animal's foot)—and Mademoiselle Romine, my family name. One English teacher who loved puns remarked it was Mademoiselle Romine, as in "You row your boat and I'll row mine!" We had the typical French two-hour lunch break and taught until late afternoon. Some of us had lunch together. I explained about my experiences at the Sorbonne. They said, "No one is checking up on you. Travel around and get to know France."

Allure of the Open Road—An Introduction to Auto-Stop

I took their advice to heart. I obtained a youth hostel card for overnights, a small backpack, and a road map of France. I would try hitchhiking, which was recommended and accepted at that time for young people with the travel bug and little money. First, I got to know Fontainebleau, its castle and history, its remarkable gardens, and the forest. The building in which I was teaching had been the home of a royal mistress. The room where I taught had a full-length gilded mirror left from the eighteenth century. It was a bit disconcerting to see my reflection while teaching!

One autumn Thursday after my last morning class, I took my little pack and walked along a street heading out of town to a road with an arrow indicating Chartres to the left. I was beginning my tour of France. It never occurred to me I might be stranded in the middle of nowhere or that I might find myself in dangerous situations. I was an American innocent abroad. We were numerous in Europe in those days.

I had studied the Chartres Cathedral in my humanities class at Colorado University. I was eager to see it for myself, but first I had to get there. How did one hitchhike in France, or auto-stop, as they called it? I stood on a broad grassy shoulder of the road. As cars approached, I raised my hand. The drivers looked at me and drove on past. After about five minutes, a small, blue-grey Deux Chevaux, a car not unlike a large, corrugated tin can on wheels, bounced to a stop in front of me. A matronly lady in a blue housecoat reached across the front seat and called in French, "Do you want a ride to Étampes? I'm going to the market, hop in if you do." I did. Étampes was halfway to Chartres. It was my first time in a Deux Chevaux, with its roll-down fabric roof and its loose suspension. It was fun, not fast. The woman right away noticed I had an accent in French. She asked if I was German. No. Was I English? No. Was I American? Yes.

"Oh, good," she said in French. "I don't speak English. My husband works on the American base and almost no Americans speak French. It is so good to be able to talk with an American. Isn't it terrible what is happening in Berlin with the wall going up? That will keep the Americans stationed in Europe." Right away I knew auto-stop would be a better education for me than sitting in a French classroom. I struggled to discuss politics in French, but I was excited to be communicating in a mobile classroom with my personal, albeit unknowing, tutor. After about thirty minutes, the informal lesson ended and she let me out on the road to Chartres.

It was another half an hour before I had my next ride. This time it was with a pharmacist from Chartres. He was proud of his city and cathedral. He said, "Keep looking out over the fields there. Soon, you will see the cathedral." Sure enough, two unmatched towers rose above the plains in the distance. They grew higher and more impressive the closer we came. Soon the cathedral dominated the view. It was splendid! So much more than the slides in my college classroom! My driver took me right to the cathedral with instructions to take my time, to see it at different times of the day, to pay attention to the old stones and the stained-glass windows. Thanking him for the ride and instructions, I slid out and said "Au revoir, monsieur." I had hitchhiked to Chartres!

The cathedral façade and the central tympanum of Christ and the four gospels rose above me. Details from a college lecture came alive in these twelfth-century figures. When I entered the nave, all the facts seemed to disappear into the wonder of a sacred space. I walked beside the mighty pillars, which were like trunks of a holy forest rising to the high vaults above and stood in silence below the translucent blues of the stained-glass windows. Their symbolism, however significant, seemed diminished by the sheer beauty of such ethereal light. In Chartres, I discovered a dimension to life I had never felt before, as history, art, architecture, faith, gratitude, and awe came together in one vast edifice.

The tinkle of a bell broke the silence. It was the verger announcing the closing of the cathedral for the day. I walked slowly out of the darkening cathedral into the evening streets. I found the youth hostel, and so ended my first day on the open road.

More Alluring Auto-Stops

Every time I had a chance, I took off hitchhiking. It was enticing! One ride was with a true French royalist. He was campaigning for the Count of Paris to become the head of state in France. During another ride, a devout Catholic took me to the home of Sainte Thérèse of Lisieux. There, at a simple shrine, he said a prayer before continuing our drive to Rouen and its famous cathedral.

I particularly remember a fascinating ride with a French educational specialist who had just returned from a remote area of the Caucasus Mountains in the USSR. He had been hired by a Russian-backed warlord to create a school system. He described flying into a landing strip, far from any town. Men on horseback with bushy beards and turbans met the plane with its three educators. They all unloaded school supplies and Russian books and packed them on horses. Then other men in turbans arrived with horses for the educators. They had not been told about having to ride on horseback. An interpreter arrived last and clarified the details about their two-day ride into the dusty village of the warlord. They were welcomed with true desert hospitality. Women brought them food and disappeared silently into back rooms. Women and girls had no part in this school. It was for boys and men to become literate and learn basic skills and knowledge. My driver wondered how it would go. That was over sixty years ago. Things seem not to have changed very much, especially when considering the Taliban treatment of women in Afghanistan today.

One weekend I decided to go to Geneva. After several rides toward the Swiss border, a large black limousine stopped for me. Inside were two other hitchhikers. The driver, who looked North African, explained that he was Muslim. As such, he believed he should help those in need. He was driving to Geneva, which meant he should take others with him since he had a large car. It was part of his faith. We expressed our gratitude and discussed many things with him, including making sure we had our passports for the border.

Then, as the automobile came up over the Jura Mountains, we saw Lake Geneva and Mont Blanc, fifteen thousand feet high, in all its glacial glory, right in front of us. Our driver said it was time for one of his daily prayers. We wouldn't need to stop; he had a tape of the call to prayer, so we would hear it as we drove along. Soon a prayer was echoing through the car, recorded from a distant minaret on a faraway mosque. One driver and three riders from as many different backgrounds were silent in the splendor of the view and the reminder of the holy dimension. Later, he drove us right to the youth hostel.

On the Road to Avignon: An Incredible Twist

On one long weekend, I decided to hitchhike to Avignon in southern France. I had just received a letter from Pop reminding me to be careful of the French. I knew he disliked the French and didn't trust them. His negative attitude started during World War II when he was working with the French allies in an Italian prisoner-of-war camp in North Africa. Pop oversaw the food supplies in the camp and was often missing certain items reserved for the Italian prisoners. He could never catch the culprits, but he knew they had to be French, as they were the only other guards with access to the supply tent.

Fast-forward to France some twenty years later. I was standing by the road that headed south through the forest of Fontainebleau. A big Citroën driven by a middle-aged man stopped for me. I indicated I hoped to go to Avignon. The man, with an accent of the Midi, said he was headed for Spain. He would be driving right through Avignon and would be glad for the company. I hopped in and we drove off.

The trip was pleasant. We spoke of many things, and he seemed nice enough. He even invited me for lunch in a Routiers, the truck drivers' restaurant chain in France. After lunch, the conversation turned to food when he said, "You know, mademoiselle, I was in North Africa during the war. I learned to love couscous there during those days. I was in the French military, working with the Americans in guarding Italian prisoners of war. The Italians were happy not to be fighting in the war. It was an easy tour of duty for me. Besides, my buddies and I often ate better than the folks in France with rationing. I was able to get into the American supply tents and find food for special occasions. I really had to be creative not to get caught."

I couldn't believe my ears. Could this kind man, trying to help his comrades party well, have been stealing Pop's supplies? He had his reasons, so I didn't say anything.

Later, when I saw Pop, I told him the story. His response: "I told you so!"

Auto-Stop in Brittany

Other trips by auto-stop took me to the province of Brittany, the far western part of France. I visited Mont Saint-Michel, the picturesque tidal island with its grand abbey and medieval-looking shops and buildings. My first trip, in February 1962, was like a step back into the Middle Ages. My auto-stop ended on the coast. My driver said he didn't know what I would find in winter. I had to walk a cold, windy mile across a primitive roadway to the island itself. During periods of high tide, this site was cut off from the mainland, but decades ago a causeway was built to take tourists right to the town.

Flocks of sheep grazed on the surrounding pastures as I approached the island. I could see a shepherd with his heavy cloak buffeted by the wind as he leaned against his wooden staff. I was the only tourist around. There wasn't a youth hostel, but one family had a spare bedroom where I could stay overnight at a minimal price. I was free to wander around at will. As I walked up the cobblestone street to the Romanesque-Gothic abbey, I could hear the banging echo of the builders renovating the edifice. Later I heard male voices singing the evening vespers. I felt transported back to the Middle Ages. I had a simple supper of the local omelet specialty and fell asleep absorbed in this medieval atmosphere. Today, I see video productions of Mont Saint-Michel showing dozens of huge tourist buses disgorging crowds of visitors into jammed streets and noisy chapels. I think how fortunate I was to have seen it when and how I did.

It was difficult to hitchhike within Brittany. I had to take small local roads to get where I wanted to go, which was to prehistoric sites. There were numerous Neolithic monuments around Carnac and the Gulf of Morbihan, dating from around 3000 BC. There was nothing as dramatic as Stonehenge in England, but there were hundreds of standing stones, some placed in lines across pastures surrounded by grazing cows.

One captivating ride was with an oyster fisherman, an amateur archaeologist himself. He drove me to a site where a sixty-foot-high menhir, a prehistoric stone originally placed upright, once stood. It had since fallen, breaking into four pieces—the remains of the tallest menhir in Europe. He showed me his oysters, which were seeded underwater on roof tiles. They were then replanted several times as they grew over five years until they were ready to harvest. I wondered if Neolithic people had an oyster diet. He said, "Probably, since many ancient people lived on a variety of shellfish."

He drove me to a dolmen, a huge, flat, table-like stone on top of several other boulders that served as table legs. Many, if not all, of these were originally covered by soil and were tombs of important people. He said the best dolmen was on Gavrinis, but the island was impossible to reach without a boat. I thanked the oyster fisherman for his ride and for taking me to the stone sites. He said he was always fascinated by the stones. I decided to go across the bay to the Gavrinis dolmen tomb.

Boat Hitchhiking, or Bateau-Stop

I could see a large grass-covered mound on an island out in the Gulf of Morbihan. How to get there? I came upon a teenage boy puttering around in an old motorboat. I called out greetings and asked what he was doing.

"*Pas grand-chose!*" (Not much!)

"Have you ever taken a hitchhiker on a boat ride?" I asked.

"Not until today," he replied. "Do you want to go somewhere?"

"I hear that the Gavrinis Tumulus is not to be missed," I said.

"*Eh bien*, it is locked up, but I know how to break in, so if you are interested, let's go, *allons-y!*"

I sat in the bow of the dilapidated boat (at least it wasn't leaking) as the craft putt, putt, putted us out to the small island, perhaps a kilometer offshore. A huge moss-covered mound rose back from the shore. The boy led me around to an iron door with a large, rusted padlock. He took out his pocketknife. He inserted the blade in the lock, which clicked open. He said that he and his buddies liked to come out here and explore the tomb. They never disturbed anything, and no one seemed to mind.

He reached up on a ledge inside the door and found two candles and a box of matches. By candlelight, we explored the center aisle lined by great rock slabs. Each slab was decorated with spirals ground into the surfaces. The boy who had been saying, "Here, look at this. See that!" suddenly became quiet. The mysterious spiral forms seemed to move and turn in the flickering candlelight. We were transported back thousands of years to the burial of an important figure, laid to rest on a stone bed in the back of the tomb. I would have stayed longer, tracing the spirals with my finger, soaking in the atmosphere. The boy said that he should go home. We made our way back to the door, extinguished the candles, replaced them on the ledge, closed the door, and relocked the lock. We were back in 1962. The boat started right up, and we headed to shore. I gave the boy a few francs, which he reluctantly accepted. I said, "For the fuel." He said he liked to go to Gavrinis, but not alone. I had done him a favor. So concluded my first and (so far) only bateau-stop!

The Way to Luxembourg

In the summer of 1962, my sister Sue came to visit me and flew into Luxembourg. I told her I would meet her at the airport. Getting there was a challenge. I had to hitchhike to Paris, take the subway across the city to the north side, and then hitchhike further north to Luxembourg. All went well until I came out of the subway at Porte de Pantin. It really wasn't on the route out of Paris, so I walked on to the outskirts. By then it was about noon, and most Parisians were inside eating their main meal of the day. The traffic leaving town had disappeared. I had eaten breakfast hours before and was getting hungry, but I didn't dare step into a restaurant for lunch or I might miss a ride to pick up my sister. To make matters worse, the bakeries had closed for the lunch break.

I continued, slowly walking north. No food in sight. No rides. Finally, shortly before 2:00 p.m., lunchtime concluded and the traffic picked up. I got serious about catching a ride. I knew it couldn't be in a Deux Chevaux. It would be too slow and likely just a short ride. I only signaled large cars. One stopped. The driver, a middle-aged man, asked where I was going. "Luxembourg or anywhere on the route there." He said he was headed to Luxembourg but would have to make a short detour on the way. He needed to get to Luxembourg before the offices closed at 6:00 p.m. Sue's flight was landing at 7:30 p.m. It might be close, but possible. I thanked him and slipped into his car.

My driver was an architect from Paris. He had a client in Champagne, the area we were entering north of Paris. He had designed renovations for a cellar and needed to check on details with the owner. We drove on the main highway for a time and then turned off onto a small road that meandered through the vineyards. It was very picturesque and a welcome diversion. We turned off into an even smaller alley and then into a courtyard. A barking dog trotted up to the car, but stopped when his owner came out and greeted my driver.

They went through the normal French niceties of extensive handshaking, and I was introduced as a hitchhiker. This was the owner of the champagne cellar, and I was invited along on the tour of the cellar and the remodelling work. But before we began the tour, we had to have a bit of champagne. I had never drunk real champagne that was produced in the Champagne province. I had had nothing to drink or eat since breakfast, so of course it was wonderful! Before the tour was over, we had enjoyed three glasses of the best champagne! I'm sure I was giddy and silly, though I tried to act sober. I don't remember much about the rest of our trip to Luxembourg, but I won't forget my first real champagne! I thanked my driver profusely as he let me out near the airport. I was there with time to spare—time to get a bite to eat before I met my sister and to sober up a bit!

A Greek Odyssey

Prologue

Any chronicle of my life would be incomplete without including the huge role of my husband, Hajo. We met initially in Denver in 1961 while teaching at George Washington High School. We arranged to meet again later that year in Europe and "dated" our way around Scandinavia. Hajo had grown up in wartime Germany. The threads of his life there, his relationships and experiences, became intricately woven with my own. They compose our odyssey in which wandering and survival often took epic proportions, especially in his youth. They directly influenced our wedding in Germany, our honeymoon, and later adventures in Greece.

In 1944, when I was an elementary school girl peacefully exploring the mountains around Boulder, Pop was somewhere in North Africa in a war I didn't understand. At the same time, Hajo, a twelve-year-old pre–Hitler Youth schoolboy, lived in Berlin, a city under siege. One morning, after sleeping in his home cement shelter as bombs rained down overnight on his neighborhood, he awoke to discover that his school was no more. His class was immediately evacuated to a village in Poland for safety's sake. He went with his school buddy Dietrich Altmann, among other students. They were accompanied by an elderly schoolmaster, a teacher of ancient Greek, too old to fight in the war. In the village, they were often conscripted to work in the fields. Their school lessons centered on memorizing ancient Greek, which bored Hajo and Dietrich but, despite everything, inspired a love for Greece.

Dietrich's father, a Lutheran pastor, was part of the Confessing Church, the anti-Hitler movement from which came the well-known theologian Dietrich Bonhoeffer, who was arrested and executed by the Nazis. (His life and sacrifice continued to inspire Hajo, me, and so many others to stand up against dictators, fascism, and totalitarianism.) But, back to our story.

In late 1944, as the Russian army advanced toward Poland, another enemy struck Hajo's class: a diphtheria epidemic. Without drugs and vaccinations, almost half of the class died of the disease: the tragic irony of dying from bacteria while escaping bombs! Both Hajo and Dietrich survived the scourges. But the Russians continued their advance.

Hajo's father, who had been severely injured in World War I, was excused from fighting in the Wehrmacht. Following the news of the Russian advances, he was determined to get Hajo out of Poland. Somehow, he convinced the authorities that Hajo's mother was deathly ill and wanted to see her only son. With official papers in hand, Hajo's father arrived in the Polish village with two bicycles. He and Hajo biked back to Germany.

At one point, American planes strafed the narrow road on which they were riding. Father and son saw them approaching and flung themselves into an empty ditch for cover. The planes roared past overhead as exploding bullets kicked up dust along the road beside them. They survived unscathed.

The usual route to the part of Germany where the family was to be reunited led directly through Dresden. At that time, Dresden had not been destroyed, because it was not a strategic city. Even so, Hajo's father said they would avoid Dresden because one never knew. His caution was rewarded. From February 13–15 of 1945, Dresden was pounded by over one thousand bombers of the United States and England. Incendiary and heavy bombs left Dresden in shambles, with over twenty-five thousand deaths. On February 14, Hajo and his father crossed the Elbe River thirty miles south of Dresden. They reported cycling through a storm of ash falling from the firebombing of the city they had avoided. They were once again survivors of the relentless violence of war.

The end of World War II brought a period of reconstruction and reeducation. High school students in Germany were encouraged to apply for exchange programs to the United States. Hajo was chosen for a family in Gloucester, Massachusetts. His exchange father was himself an immigrant, a Greek sculptor. Hajo's life with the Demitrios family cemented his love for Greece and his determination to immigrate to the United States. Meanwhile, Rolf Peterson, Hajo's uncle and sometimes mentor, became head of the Goethe Institute in Patras, Greece. Back in Berlin, his friend Dietrich Altmann had become a pastor like his father.

By the summer of 1963, Hajo and I had become officially engaged. After meeting Hajo in Denver two years earlier and dating across northern Europe for over a year, we planned our wedding to be in Germany. We chose the Lichterfelde village chapel in Berlin, which had survived the bombing. However, the day before our wedding, a violent thunderstorm blasted through the area, with a lightning bolt blowing out the shutters of the chapel bell tower! Was this a bad omen? The wedding continued in any case, officiated by Dietrich Altmann. I felt our wedding was not only a beautiful gathering to celebrate our union but, in some small way, part of the Confessing Church's values. Hajo's uncle Rolf came from Greece to be one of our official witnesses. Our families and friends attended the celebration and then sent us off on our honeymoon . . . to Greece!

We drove south from Berlin in Hajo's secondhand Volkswagen Beetle. It was old, but dependable. We called it Fancy, not because it was ornate, but because it took us "wherever Fancy took us!" We were loaded down with essential camping gear and a large carton of C rations: packaged meals American soldiers carried for emergencies. Hajo had collected them during a short stint of US military service several weeks earlier. He had served with an army publicity crew covering President Kennedy's visit to Berlin. He'd stood on a press truck near the president as he gave his "Ich bin ein Berliner" speech to the cheering crowds. Hajo loved C rations because, as a constantly hungry teenager right after the war, he had received them from kind American soldiers. C rations, tent, Primus stove, bedding, and don't forget the guitar on the top of the pile. I knew how to strum a few chords, and we liked to sing. We were off on our odyssey.

En Route

We headed south through Bavaria and Austria into Yugoslavia. When we arrived in Belgrade, we began our search for a campground. In those days, every car required a decal with the country of origin. We had a USA sticker. A big black sedan drove around us with a CD decal. "What country was that?" I wondered as it flashed its lights for us to stop. It was a Corps Diplomatic sedan. We pulled over and a gentleman in perfect English greeted us, saying he was with the US Embassy, and wondered if he could be of help. We replied, "It's great that our embassy is watching out for us!"

He explained, "I saw your USA sticker. It is so rare to see Americans driving here we that we decided to stop you."

"We're looking for a campground," we continued.

"I don't have much time for camping, but I'll ask the driver." The chauffeur gave us clear directions, and we thanked them as they drove away.

As we watched their car disappear around a curve, Hajo suddenly exclaimed, "Oh wow! It just hit me. I should have recognized that man. He looked familiar. I am just now reading his book and saw his picture on the jacket. That was George Kennan, the American ambassador to Yugoslavia. Imagine the ambassador stopping to help us. That is the true spirit of America!"

The next day, I think it was July 11, we continued driving south. We camped that night in Skopje, Macedonia, then still part of Yugoslavia. The campground was nestled down in a V-shaped valley. I have a photo of me cooking near our little tent. If we had been there two weeks later, I would not be writing about this. On July 26, 1963, a magnitude 7 earthquake shook Skopje. The epicenter was the campground. Over one thousand people lost their lives and two hundred thousand were rendered homeless. Most of the city was destroyed. We drove by Skopje on our way back north in August and passed hundreds of people wandering aimlessly before the international community came to their rescue.

Discovering Greece

After a few days of driving and camping, we reached our goal: Greece! Land of architectural masterpieces and ruins, dramatic seascapes, beaches, and the blue Mediterranean, of Greek dramas, statues, the Olympics, and so much more. We took in these magnificent and fascinating sites, the stories of which fill the tourist guides and history books. It was all that for sure, but I will comment on another Greece we met in the process of our camping trip.

We had planned to climb Mount Olympus, but an overnight rainstorm flooded us out of our campsite and turned the trail up Mount Olympus into a river of mud! Totally unpassable! Probably a good thing because it turns out the monastery on the summit did not allow women to spend the night. The next evening, we found ourselves at Meteora, the rock formations and pillars with Eastern Orthodox monasteries perched on top. We splurged on a restaurant supper so we could camp in a spectacular setting under a tree behind the restaurant kitchen. We fell asleep with the tent door open and the odor of delicious Greek spices filling our dreams.

In the middle of the night, something woke me. I peered out the tent door to a landscape filled with moonlight and dramatic geological formations. But then I noticed that the land in front of our tent seemed to be moving. I quickly put on my glasses for closer investigation. A wide trail of ants was marching from the woods behind our tent to the restaurant kitchen. Their route was several inches from our tent door. I roused Hajo. We zipped our tent closed and stuffed tissues in all the cracks. No ants had entered our tent: they were "intent" on going for scraps in the restaurant! We fell back asleep. The next morning, there were no indications of army ants.

At some point, we stopped at a roadside vegetable stand. A Greek salad seemed just right for lunch. I pointed out to the Greek owner the tomatoes, green peppers, olives, and spices we needed. But there were no onions. I couldn't find the word for onions in my Greek vocabulary booklet, and Hajo's ancient Greek didn't help. We tried German and French. I showed a round onion shape with my hand, but the owner just looked perplexed. Then I started with fake tears to indicate what an onion could do. The owner put his arm around me to comfort me. I started laughing and the light bulb suddenly flashed over his head. He hurried to the back room and returned with two beautiful white onions. I had already paid for my other purchases and tried to pay for the onions. He refused. "No, no! Souvenir, souvenir!" he said with a big smile. A Greek salad never tasted so good!

Somewhere on the Peloponnesus, we camped off the roadside near a small beach. It was lovely and lonely. Following an afternoon swim and a supper of C rations, I picked up my guitar and started strumming and humming. Hajo was at work on something by the VW when some guys in a Greek car drove up, jammed on the brakes, and stopped. Hajo pulled himself up to look strong and in charge, fingering his pocketknife as they came over. There were four of them and two of us. We didn't know what to expect.

They seemed friendly as they greeted us. We responded. One of them pointed to my guitar. I hesitated but gave it to him. He started playing and the guys began singing. We all sat around on rocks and the camping tarp. For an hour they sang Greek songs for us, and we sang American songs for them. They especially liked the spirituals, like "Swing Low, Sweet Chariot." After a while, they stood, returned the guitar to me, and sang a parting song. They piled into their car, all smiles, and drove away, leaving an unforgettable memory.

One of the spirituals came back to me later that week. It was early morning as I peered out of our tent and gazed along the wide sandy bay where we were camped. Around on the other side I could see a small Greek chapel. The sand was smooth and inviting. I jogged and splashed my way around the bay to the chapel. The door was open, but no one was around. I looked inside. There was a small white altar and beside it a simple statue of the Virgin with the young child Jesus in her arms. He was holding what I supposed was an orb, a symbol of royalty, but this was a light-blue ball, like Earth seen from space, being held gently by the Holy Child. "He has the whole world in his hands / He has the whole world in his hands / He has the whole world in his hands / The whole world in his hands." The spiritual stayed with me as I strolled back around the bay to our campsite.

Another time, Hajo and I were hiking down a dusty path that led by a small chapel. As we approached, we could hear singing. Then the door opened, and a few people left. One woman, all in black, walked toward us and gave me a small branch of basil. She smiled and said a few words we couldn't understand. Perhaps a blessing? We thanked her. It was such a simple gesture, but the fresh green branch and leaves were so in contrast to the harsh, dry, stickery vegetation around the trail, and her smile so warm and kind, that they were the message more than her words.

Before we left Greece, we made a pilgrimage to the Acropolis. In high school in Boulder, I had created a model of it with matchstick pillars for the Parthenon and attempts at other statues. As primitive as my model was, I knew the Acropolis very well. I was thrilled to see in real life this cultural creation that was more wondrous than I had ever imagined! What an appropriate way to end our stay in Greece. We drove north from Athens and back to Germany, our honeymoon over. We returned by freighter to the United States, drove cross-country in Fancy, and began our married life in Colorado.

Return to Greece

The memories of Greek columns, customs, cuisine, and camping stayed with us over the years. In 1988, twenty-five years later, we returned to Greece. This time, Hajo was teaching international relations to US Air Force officers stationed near Athens and on Crete. Our son Eric and daughter-in-law joined us for their own honeymoon in Greece. They have their own memories of their week with us. I was along for a three-month celebration of early retirement from teaching in Montana. Our home near Athens was next to the airport runway, with landings and takeoffs our daily music. Fortunately, at night no flights were permitted, and in the silence, we inhaled the fragrance of the jasmine bushes under our windows.

During our free time, we read *Eleni*, the story of the courageous Greek mother who saved her children from deportation from Greece to Communist countries during the Greek Civil War in the late 1940s. Nicholas Gage, an American journalist of Greek ancestry, had written this captivating story about his mother and sisters. His mother had been executed by the Communists for sending her daughters and Nicholas to the United States rather than having them sent elsewhere for a Communist education. He had wanted to know what really happened to his mother in this war of competing powers. Near the beginning of his story, he quotes an old Greek proverb: "When buffalo battle in the marsh; it's the frogs that pay."

Hajo and I had just finished reading this book when Hajo had a free long weekend. We decided to go and see for ourselves where this story had taken place. We packed our camping gear in the trunk of Hajo's dream car, a large white Mercedes—the kind used for taxis in Berlin, the workhorses of the Mercedes line. Hajo loved it like a delicate mistress and polished and protected it from every bump or scratch.

Our tour took us to the Greek–Albanian border. The Communists were still in power in Albania, and we were very aware of the tension across the border. We reread parts of the book as we explored the village roads, rundown homes, and farms, and relived the tragic death of Eleni. We considered going into Albania, but decided it was too risky. As I write this now, two generations later, the Russian Communists are now attacking Ukraine, and the frogs are still having to pay.

At the time of our visit to this region, back in northwestern Greece, farmers were having a major dispute with the Greek government—if I recall correctly, over water rights. The government was threatening to divert irrigation water from farmers' crops for other projects. The farmers were desperate because they depended on this water for their livelihood. They didn't have much leverage for their cause, but they did what they could, which was to blockade all the roads in and out of the northwest area of Greece. We had heard of possible unrest but didn't imagine it would concern us.

Sunday morning, we took down our tent, packed everything in the Mercedes, and drove to the highway that would take us back to Athens for Hajo's classes the next day. The highway was open until suddenly there was a major traffic jam. We stopped in the row of stalled cars. Everyone was waiting patiently, but there was no traffic moving in any direction. The American Forces News Station came on our car radio. There were rumors of traffic disruptions in the northwest of Greece, and Americans were warned to avoid travelling in that area. Exactly where we were! Hajo really needed to be back in Athens for his classes. We had twenty-four hours. Surely, the disruption wouldn't last that long. And there must be other highways.

I stepped out of the car to survey the situation while Hajo checked a road map of Greece. There were dozens of vehicles ahead of us, mostly Greek cars mixed in with huge tractors and farm wagons blockading the highway. Then I spied a

French car. Maybe the driver would know what to suggest. At least I could communicate with him.

He and his French wife were vacationing with distant Greek relatives whom they had come to visit. Their elderly relatives were in the car in front of them. They said the blockade was scheduled to last until the government gave in to the water rights of the farmers. That would take at least a few days. But the couple said their relatives knew of a wagon track that could take them another way around the blockade to this same highway further on. They said it was very rough but probably passable. We could follow them if we wished. They were going to turn around at any time.

I hurried back to Hajo with the news. He had not found any alternatives on his map. "Let's go!" he said. "I have to get back to teach or I risk losing my teaching contract."

All three cars circled back. We came to an unmarked road that led us through several rural farmyards. Chickens ran away from our caravan. We stopped several times to open half-unhinged wooden gates and drive through as the road became narrower and rougher. Then two tire tracks cut across a meadow toward a hill of rocks, a talus slope with a slanted narrow track across it. The Greek relatives left us there.

The French couple said all we had to do was cross the rocks and the road would open up and lead us to the highway we needed. Hajo wondered what we had gotten ourselves into. Could we get his Mercedes across unscathed? The French woman and I got out of our cars and walked ahead of them, clearing the rocks to widen and smooth the tracks and make them passable. Our husbands maneuvered gingerly behind us. We made it without a scratch. We expressed our gratitude and adieux to the French couple. Soon we were back on the highway, driving to Athens. Traffic was normal, and we arrived at our Athens home as though nothing had happened.

The next morning, the news announced that northwestern Greece was still blockaded and that some one hundred tourists were stranded in the area. The blockade continued for several weeks with no known ways in or out. When Hajo finished his term in Greece, the farmers and the government still had not reached a settlement. This particular blockade eventually ended, but others were used to paralyze traffic in Greece for various causes over the years.

While Hajo was teaching, I had time for cultural events. The Roman amphitheater at the foot of the Acropolis was a venue for many concerts. I heard that the Bolshoi Ballet was scheduled to perform Prokofiev's *Romeo and Juliette*. Hajo was teaching that night, but I tried to get a ticket. Totally sold out! A friend wrote me a sign in Greek: "I am looking for a ticket." The night of the performance, I stood outside the entrance with my sign. People looked at it, shook their heads, and entered. Then some young Greek woman stopped and said that a friend had an emergency and couldn't attend. Would I like to buy her ticket? Indeed, I did!

It was a deeply moving experience: the juxtaposition of ancient drama in this theater with the Romeo and Juliette story from Shakespeare reworked into ballet. In front of me, the stage was lit up by stage lights, but a full moon also illuminated the Acropolis beside and behind us. It was a dramatic setting for an intense, universal human drama. The Prokofiev score and the dancers, as well as the two groups of warring families, built in intensity as love and impending death unfolded on the stage. Suddenly, in the audience right before me, an elderly man collapsed into the lap of an old woman, who called out in alarm. People surrounded them, trying to give the man lifesaving care until a medical crew arrived with a stretcher. The man was covered over without a sign of life, and his weeping wife was aided out behind him. On the stage, the ballet continued without a pause as love, conflict, tragedy, death, and grief played out. As I exited the ancient amphitheater, a shiver went through me as I reflected on the historic setting where two thousand years of such drama had unfolded, both on the stage and in real life.

Crete

Crete is a remarkable island off the southern coast of the Greek mainland. It is the site of the Minoan culture, the oldest civilization in Europe, which began about 3500 BC. It had trading relations with Egypt, but a different culture and architecture. The restored Palace of Knossos, with its sophisticated frescoes, displayed Minoan civilization at its height. It can be visited and admired today, as can other sites. However, they all suffered destruction, probably by earthquakes or perhaps by the eruption of Santorini (a nearby volcanic island), known today for its stark white villages set against the blue Mediterranean. Hajo and I visited Crete and Santorini several times, but I am not writing a tourist guide, just setting the stage for our visit to a geological wonder of Crete.

Less well known than Knossos, the Gorge of Samaria is a national park in the western part of Crete. Its highlight is a ten-mile trail cutting through Crete north to south from the four-thousand-foot height of the trailhead down to the Libyan Sea. Hajo and I took a local public bus to the trailhead, not knowing if the trail was open, because it closes at the first sign of an impending rainstorm. It was open. We read the warning in multiple languages: "Narrow passages ahead with danger of flash floods and no escape routes!"

The sky was cloudless as we proceeded down the trail. We could see rough stretches of broken landscape in front of us, but no real gorge. We met a sprinkling of other hikers. Wildflowers dotted the meadows along the sunny trail. It was easy going gradually downhill. Then the trail became steeper as we cut down a jagged valley. We stopped at a viewpoint and talked to some Greek hikers. They had lost a friend in a flash flood here several years earlier. They were on a memorial hike. We kept our eyes on the sky. No clouds.

The trail dropped down, down, as the canyon walls rose around us. Then the trail levelled out. We had reached the bottom of the gorge. A trickle of a stream meandered down the gradually sloping riverbed. The trail disappeared into a rugged jumble of scattered rocks and boulders. We followed the dry bed until we were encased by the canyon cliffs. Called the Iron Gates, the thousand-foot vertical walls rose straight up on all sides of us, forming a thirteen-foot-wide canyon. Hajo and I together could reach across it. No sun penetrated this shadowed realm, grand and mysterious! A rainstorm ten miles up the canyon could send a raging flood roaring through this narrow channel despite perfectly dry conditions where we were. Fortunately, it was a dry, still day all over Crete!

We slowly clambered over rocks and rubble as we moved along and then out of this canyon. The vertical cliffs dissipated into rounded hills, and soon we were hiking in hot barren desert. We wondered if the people of the Minoan world had ever ventured this way. It seemed likely, but nothing indicated or denied their presence there before us. We rounded a hill, and there in front of us stretched the Libyan Sea. The sun was brutally hot, the sea blue and inviting. As we drew near, we dropped our packs, kicked off our boots, and threw ourselves into the refreshing waters. The waves washed over us as we laughed and sprayed each other, a gleeful conclusion to the Gorges of Samaria. We sat in the sun drying off and then followed a goat track along the seacoast to a bus stop and our ride back to civilization. What an amazing trek! No wonder the Gorge of Samaria is known as the European Grand Canyon!

Crete | 73

Figure 11. Deux Chevaux: The slow way to go!

Figure 12. The Lequints: My Swiss Farming Family

Figure 13. Hajo and I begin 53 years of adventures together

Figure 14. Fancy, the VW Beetle

Figure 15. Camping in Skopje shortly before the devasting earthquake

Figure 16. One of many family trips to the Matterhorn, the mountain that inspired my travels

"O Give Me a Home..."

Prairie Montana

Hajo, our two sons, and I lived for twenty years in Havre, Montana, originally known as Bullhook Bottoms. Havre is located on the northernmost rail line across the western United States. US Highway 2, on which we drove into town for the very first time, parallels the rail line. Canada is twenty miles north of where Havre's venerable Corner Bar stood at the time. A sign even said so ("Where's Canada? 20 miles north of the Corner Bar")! Up north, the Alberta and Saskatchewan borders meet.

When we drove into Havre, population ten thousand in 1967, I felt like we had landed in the dingiest burg in the West: dusty streets, tumbleweeds strewn about, grain elevators, the clanking of trains being shunted around, and no mountains in sight! Hajo, who had been there earlier for an interview, said that it would grow on me, and I tried to reserve my judgment. The nearest city of any size was Great Falls, population eighty thousand, exactly 111 miles to the next stoplight southwest of Havre. We pulled a U-Haul trailer from Denver with all our earthly possessions into town from the southeast. We had driven for three days through wheat fields and native reservations, across the Missouri River, and through broad stretches of wild prairie lands. Our son Eric, age six months, slept most of the way in a cardboard-box travel bed.

Hajo was driving to his first teaching job since completing his doctorate in international relations at Denver University. He was a professor at Northern Montana College, an outpost of education for farmers, ranchers, railroad workers, inhabitants of dozens of small towns, Native Americans, and hospital employees. The citizens of hundreds of square miles had only this public college available in the wilds of North

Central Montana. I was considered his dependent wife, an identity I resented, because despite my credentials, letters of reference, and years of teaching experience, it was six years before I was hired as a full-time teacher in Havre. I learned a valuable lesson in what it means to be unemployed!

The weather in Havre was an ever-present challenge: wind, almost constant wind, with blizzards and bitter cold and ice-covered roads in winter. Occasional warming Chinook winds blew in from the southwest, and the temperatures could jump from minus twenty degrees to thirty-five above. Such temperature spikes inspired one driver of a convertible to open the roof and drive through town to celebrate the "heat waves." Our first home, a small frame house (one of several furnished to NMC professors at minimal cost) had little insulation and scarcely protected us from the elements. Most of the first winter in Havre, any water dropped on the kitchen floor froze, along with canned goods in the flimsy cupboards. Eric spent much of his first year in his crib three feet off the floor because the air temperature was thirty degrees colder on the floor than on the ceiling.

Summers presented the opposite challenge. The wind and ninety-plus-degree temperatures created a blast-furnace effect that choked the life out of every garden I attempted to grow. These were some of the physical challenges during the first years, before we could afford to purchase a real house. But through it all, my inner attitude of disappointment at our situation began to change. We were still living in an area where military personnel at a nearby base received extra hardship pay for their assignment, but I experienced a revelation that inspired me through our challenges.

Shortly after our arrival in early September in Havre, we drove on a dusty road to Fresno Reservoir, an irrigation lake on the Milk River, to have a look around. A few large, scraggly old cottonwood trees sprawled beside a rough-textured sandy beach. The area was deserted except for a few fishermen in small boats way out on the lake. The water was murky, but it seemed clean, so I went for a swim while Hajo watched Eric, who was asleep in his box.

I stopped swimming a short distance from the shore and stood up to see how deep the water was. Up to my waist. I took a quick step to continue my swim and immediately stubbed my toe hard! I bent over to see what I had kicked since there didn't appear to be any rocks in the vicinity. To my amazement I pulled up a dinosaur vertebra! I recognized it immediately. It couldn't have been anything else—almost a foot in diameter, and very heavy. I was so excited! Wow! That was the beginning of a new perspective. I was in dinosaur country!

Jack Horner

For much of the next twenty years, weather permitting, I was out seeking and digging dinosaur bones and related fossils. I was sometimes alone, other times with family and friends. Occasionally, I was a member of official digs, such as those I went on with Jack Horner, the *Jurassic Park* film paleontologist.

Jack was an amazing person and expert paleontologist. As a high school student in northern Montana, he won state science fair competitions for his dinosaur projects and others, but he struggled academically. He dropped out of the University of Montana, but realized at the same time he was very good at his dinosaur work. He knew he would have to gain museum experience. After multiple applications, he accepted a position at the Princeton University History Museum.

While there, he caught sight of a poster on a bulletin board. It offered to evaluate students who had difficulties reading and memorizing. Jack decided to take the required test and was diagnosed with extreme dyslexia. He said it was a great awakening! He had always been called stupid. The test affirmed that he just had another way of seeing things and was intelligent in a different way. He dedicated his life to the study of dinosaurs. He wasn't so much interested in their extinction as in how they lived and thrived for so many eons. His work included discovering dinosaur nests, eggs, and juvenile bones, and learning that some species cared for their young.

One of my friends who volunteered with him suggested I contact Jack and see if I could participate in a dig. It was January, but I immediately wrote him a letter requesting an application. He said it was too early; I should get in touch later. I wrote to him several times in the following months and never received a reply. Finally, I called him on the phone. (This was before computers.) He said he would tell me when he knew his plans. There was no official acceptance as the date grew closer. Finally, two weeks before I hoped to participate, an envelope arrived in the mail. It contained a map showing the route to the dinosaur camp and the dates in late July. Nothing else. Hajo was teaching summer school and the boys had summer activities, so I was free to go.

The date arrived. I loaded our van with a pup tent, sleeping bag and mat, clothes, sunscreen, a few food supplies, and water. I hugged my men goodbye and headed out west on US Highway 2. The Montana prairie stretched out over a hundred miles to the Rocky Mountains. Partway there, I found the gravel road taking off north where the map indicated Camp Rainier! (It was named not for the mountain but for the cases of Rainier beer that were exclusively for after-work consumption.)

I counted the miles to a wide spot on the road where I was to park. I overlooked a dry valley with a white main tent and a scattering of individual tents. No other cars or people were in sight. I thought they must be off digging somewhere. I climbed out of the van and made my way down the steep hillside, carrying my tent and all. This was obviously the right place: tools, specimens, Rainier beer. I pitched my tent near the others in the sagebrush and wild grasses. I was glad I had brought some of my own supper: cheese sandwiches and canteen water. I didn't yet know the beer protocol. By then the moon was rising over the darkening prairie. Still no one! It occurred to me I was totally alone and it might be dangerous. But what was there to be afraid of? I was the only one around. I crawled into my tent and dropped off to sleep to the distant songs of prairie coyotes.

The next morning I awoke with the sun and the song of a meadowlark. I peered out of the tent. I was still alone. It was a splendid morning. I pulled on my boots, ducked out of the tent, chugged down some water, and set off on foot to see what I could find. About a half mile out, I stumbled into a long-deserted native encampment. Clear rings of stones indicated where teepees had been pitched, with the rocks holding down skins to keep out wind and critters. These stones may have been used and reused over the centuries. Such teepee rings are a common sight on the unplowed prairies.

Nearby I discovered a chipping ground, with obsidian and flint flakes, where early hunters had knapped arrow- and spearpoints and other stone tools. I gingerly rubbed my fingers over the sharp, thin edges of the chips. I realized I was in the glacier-free zone where during the last ice age ten thousand years ago, bands of migrants had crossed the land bridge from Asia to Alaska. They had then made their way through what would eventually become Montana. At least that is one theory, and I had possible proof as I left the chips and moved on.

Further along, I found fossil animal fragments lying below a dirt cliff. They didn't appear to be anything special, but I made note of the location in case we wanted to explore further in this area. I climbed up around the bluff and discovered an entire row of petrified vertebrae, exposed to the elements and lying in a row. There I was, alone on the prairie in the present, but simultaneously transported back to early human finds and ancient fossils—all before breakfast!

About this time, I spied a line of dust: cars and pickups heading along the road to Camp Rainier. The dinosaur diggers were arriving. I hurried back to camp to meet them and tell them of my find. I knew several of the crew members, and they all welcomed me to the group. Jack Horner said they had decided to go to the county fair the day before and stayed the night. He said he was glad I had come. He explained he didn't choose people for the digs. Applications were too complicated. Participants chose themselves. If they really wanted to come and work, they would keep trying until he sent them a map.

Over the next ten days, I learned how to protect and prepare fossils to go to the museum. I dug for hours through overburden to get down to fossil-bearing layers. I learned to recognize all kinds of fossils that, up until that dig, I knew only as unidentified teeth and bones. We found lots of grey shale and discovered that mixed in it were dinosaur eggshell fragments. I could tell them by their slight, smooth curve on the inside and rough outside. Every day was a new discovery.

    This dig was a key to becoming a real dinosaur explorer. It made life on the prairie a fascinating adventure. I took my family, local children, and neighbors to sites along Fresno Reservoir, where we discovered hundreds of scattered fossils. I even found a dinosaur claw lying right on a beach where it had been overlooked for years by sunbathers.

    We camped on dinosaur sites further away from Havre and spent hours at night gazing into dark deep space, seemingly to infinity. During the day we touched deep time, finding and holding sixty-five-million-year-old fossils in our hands. The barren plains of Montana taught me to look below the surface to find all kinds of treasure. I might not be seeing the *Mona Lisa* in Paris, or climbing the Alps in Switzerland, but Havre and the surrounding area offered unique opportunities I had never imagined when we first drove into town.

It Wasn't All Dinosaurs...

Hajo and I read of the Lewis and Clark expeditions through Montana and made several float trips on the Missouri River tracing their route. Emmett Stallcop was our guide. A Havre native, Judge Stallcop was a Justice of the Peace and highly decorated for several years of military service in the Pacific during World War II. He was an expert in local anthropology and volunteered at the Havre Buffalo Jump, a local attraction. He designed and built his own Missouri River boat in the form of the landing craft on which he made three beach landings during the war. It was always a pleasure to be with him and explore northern Montana. He would call together several Havre families that had indicated an interest in going down the River, and we would go! We had to place a car at our destination, then drive back with another car to enter the river south of Great Falls at Virgil, a tiny farming community on the river.

We floated for two to four days through prairie badlands to the rare spots where a dirt road or ferry made an exit possible and where we had left a car. We were essentially incommunicado the whole time. There was a flotilla of us with canoes and kayaks around Stallcop's landing craft. On one trip we drifted past Hole in the Wall, a site identified by Lewis and Clark. On our next trip it was no longer there. Some local had decided to blow it up for no good reason. On hot sunny excursions, we would slip into the river beside our boats and drift along with them to cool off for an hour or so. On one cold, rainy trip, we huddled under tarps and drank from jugs of homemade wine that were tossed from boat to boat. We rarely saw anyone else on the river or the prairie lands through which we passed. It was genuine wilderness!

The spirit of adventure ran free on these trips, and we enjoyed hearing stories about life in early Montana. The father of one of our fellow river floaters had been on cattle drives around Montana with Charlie Russell, the well-known Western artist. She told us about how the cowhands sat around the campfire at night. Charlie would sketch caricatures of each cowboy and pass them around the circle. Everyone would have a good laugh and toss the drawings on the fire!

We sat around our own campfire. As the sparks flew into the night sky, we shared our stories of seeing the full solar eclipse that passed over northeastern Montana in the winter of 1979. Some of us had driven out early to the reservation to an isolated knoll where we had an unimpeded view of the sky. We had an extra-special dark lens on a telescope for viewing the spectacle. But a thick cloud had formed over the area, already eclipsing the sun. Disappointed, we waited in the cold, wrapped up in our parkas and boots. Slowly the cloud moved away to the east, dissipating in the process. As the moon began its passage across the sun, we could follow it clearly, all taking turns on the telescope and watching on our pinhole papers. Then, as totality approached, we watched a very dark shadow moving across the barren landscape toward us.

The last sliver of light disappeared as the moon totally covered the sun, leaving us in darkness. Our little band of watchers broke into spontaneous applause. Such an insignificant gesture for such a cosmic event! After several minutes the sun reappeared on the other side. It was like dawn. Far in the distance on a reservation farm, the crow of a rooster welcomed the new day! We wondered what the Indigenous people would have made of such an event seeing it centuries earlier. Even with all our scientific knowledge of what was happening, the eclipse left us in awe.

In Havre, Hajo and I met scores of kind, helpful people who became lifelong friends. They gave their best as teachers, doctors, hospital workers, professors, bankers, storekeepers, artists, and more. They were genuine citizens. One of my friends actually donated a kidney to a woman she scarcely knew because the woman needed one, my friend had two, and there was a match! She said it was a small sacrifice of a month of her time to give a lifetime to someone she could help!

With my family's history of hosting exchange students and Hajo's experience of being one himself, it is not surprising that we decided in the 1980s to host two high school students who became integral and ongoing members of our family in Havre. Elisabeth, from Norway, made the eyes of the local Sons of Norway members glow with pride when she attended their meetings in her traditional dress and talked with them in Norwegian. A few years later, Freddie, from Sweden, delighted her teachers when she appeared on their doorsteps early on the morning of Saint Lucia Day, December 13, wearing a candlelit crown, singing "Santa Lucia," and bringing them saffron buns. Over the years since, both Elisabeth and Freddie have hosted us in their own countries numerous times. In 2022 they met each other for the first time in Sweden when Freddie hosted a reunion. The two "sisters" hit it off right away! Such happy memories.

Hajo and I were naturally drawn to the other first-generation immigrants in Havre. I was the only native-born American in our group. We loved European-type parties with singing and dancing to accordion music played by the Havre hospital pathologist from Yugoslavia. He sang lustily with Hajo, the television repairman from Hungary, and the area meat inspector from Ukraine. The ladies from Spain, Hungary, and Germany joined in, and they all told stories of survival from the war.

During World War II, Ruth from Germany was raising her two children alone while her husband fought with the Wehrmacht. Their home was near the Polish border. With the Russians advancing in 1945, Ruth (like Hajo and his dad) wanted to get away further west. While Hajo and his dad were thirty miles south of Dresden on February 13, 1945, Ruth and her children were stranded in Dresden at the train station.

Ruth told us she had a premonition that they had to get out of the city, immediately. She didn't know why, but she had to do anything to get out!

The only trains leaving were troop trains carrying injured German soldiers. Ruth went from one train wagon to the next, begging everyone to take her and the children anywhere out of Dresden. Finally, an officer in one wagon said he would hide her and the children in his compartment under medical supplies and blankets. Ruth hoisted the children up through the windows into the train and then was pulled in herself. They would have to leave the train the first time it stopped! That was an hour later at a village somewhere west of Dresden. They were left alone on the platform as the train pulled out. That night Dresden was firebombed. Ruth's insistence (and illegal help) had saved them. Such were the stories of our European American friends!

As the years passed, some of these friends left Havre, but their children stayed in the area, becoming the next generation of Americans. They became part of the changing population as an increasing number of reservation students attended public schools and college. Hajo and I taught Chippewa Cree students and met their families. We attended many powwows on nearby reservations. We made numerous visits to the local Hutterite colonies. Several new colonies formed as the Hutterite population increased.

We led an ecumenical life, participating in church events open to all. We enjoyed our sons' school activities on the stage and sports fields. We camped, hiked, and skied in the largest county park in the United States, just fifteen miles south of Havre. We made our own entertainment with Octoberfest, New Year's Day, and Saint Patrick's Day parties, as well as high-quality stage productions, all performed with great creativity. But more than any other single activity, dinosaur exploration gave me the valuable perspective that if I keep my eyes open and am aware and present to the possibilities, life can be amazing almost anywhere.

Making the Most of Summers

One major benefit of the teaching profession is called summer vacation. That doesn't mean that Hajo and I sat around doing nothing! We both prepared lesson plans and materials for the school year. And we both had the good fortune of receiving grants for seminars. In 1985, I spent a month in Quebec and Plattsburgh, New York, on a French language program for teachers of French. All participants pledged to speak only French during our program. My roommate was a woman who had been in military service in the Pacific during World War II. She explained she was the person who sent the telegram reporting that the atomic bomb had been dropped on Hiroshima. I asked her how she felt about that. She said that she was young, it was wartime. She didn't really know what it meant at the time. I felt I was rooming with history.

We both enjoyed swimming and often swam in Lake Champlain early in the morning before our seminars started for the day. One Saturday we had no classes, and we decided to swim far down the shoreline, with plans to walk back if we were tired. After an hour or so of swimming, we headed for the shore. We found a sandy area with a couple of large rocks where we sat and dried off in the sun. Then we walked further inland, looking for a road along the coast. We came to a rail line, crossed it, and found ourselves in a neighborhood of identical houses and a street going in the right direction.

We were strolling along in our swimsuits when a jeep pulled up and a young man in fatigues jumped out. He wanted to know what we were doing. We explained we had been swimming and were headed back home. He asked to see our ID cards. We looked clueless. He asked, "Do you know you are on a military base?" We did not. "How did you get in here?" he wanted to know.

"We walked in from the lake. There were no signs or gates indicating a base." He believed us but said we would have to exit through the main gate. We climbed into the jeep and rode to the gate, and he let us out. We walked back to our campus in our swimsuits through the streets of Plattsburgh, discussing in French the ramifications of infiltrating a military base!

How about a Camping Trip in the Soviet Union?

The year was 1972, the height of the Cold War. Hajo, whose teaching specialties were Eastern Europe and the Soviet Union, got wind of a tempting possibility. The Soviet Union was inviting foreigners to come camping in Western Russia and the surrounding countries. It was a chance for Hajo to see current life in that part of the world firsthand. I was intrigued because I had written and published an article on socialist realism, *"The Case of Doctor Zhivago,"* based on Pasternak's novel and the Soviet reaction to it.

Hajo and I began the complicated process of accepting this unexpected invitation. We had to submit a day-by-day itinerary once we finally obtained a list of acceptable campgrounds. After six rejections of itineraries for various reasons and a personal visit to Intourist, the official Russian travel agency in Washington, DC, we were finally accepted for a three-week camping vacation.

We had already ordered a Volkswagen camper van to use in Western Europe. It slept our family of four, which included our two sons, six-year-old Eric and two-year-old Mark. We loaded it with peanut butter, Cheerios, crackers, and a few other staples, anticipating the limited availability in Russia of foods the boys would eat. We had in hand our passports, visas, and all kinds of other official documents that became more numerous as we were checked through East Germany, Poland, and finally Belarus.

At the first Soviet Union crossing, we drove through a channel of high cement walls with clacking metal barriers that were raised by border guards. They wore padded grey-green woolen uniforms and official hats with red stars. One guard with a rifle flung over his shoulder took our official papers and passport. He directed us to park our camper and enter a low, dark, drab building. No one else was being checked in. We sat on a grimy, rustic wooden bench, our backs against a cement wall. A single light bulb hung crookedly from a grey electrical cord in the ceiling above us. Hajo was called over to a door opened by a heavyset, uniformed woman. He entered and the door was shut. The boys and I waited outside, wondering what was happening.

When he exited, it was my turn. The same matronly-looking guard called me in. I held my breath as I was accosted by garlic breath and body odor. She gruffly patted me all over, looking for I don't know what. Then she said in broken English that she was too tired to fill out the forms in English; I would have to come back the next day. Somewhere in the waiting room, I had seen a sign in French. I responded in French to her declaration. It worked. She asked me a few routine questions in fluent French, filled out several forms, and voila, I was finished. Meanwhile, Eric and Mark sat motionless on the bench and didn't squirm or complain. They sensed that you don't fool around at borders!

We received our day-by-day "Marche Route," which listed the highways and roads we had to follow between campgrounds and where our official local guides would meet us. And no, we could not go anywhere without our local guide. We had lists of things we could photograph and visit. Other things were off-limits. We could not camp in Moscow or Leningrad because the campgrounds were full, so we were assigned hotel rooms. (In fact, the campgrounds were empty. The Russians were eager to obtain Western currency. They could charge us more for room and board in hotels than in campgrounds.)

The first changes we saw as we drove through the Soviet Union were the parallel side roads. In Poland, dirt roads ran parallel to the highways. They were used by pedestrians, and most often by horse-drawn wagons with rubber tires. In

How about a Camping Trip in the Soviet Union? | 91

Russia, these same dirt roads were more rutted, and were used by people in open wagons with iron-rimmed wooden wheels pulled by a single horse: clip-clop, clip-clop.

The routine along the roads was as expected. Several times a day, guards at highway checkpoints examined our papers to see that we were where we were supposed to be, then waved us on our way.

Each campground provided unique experiences. In most of them, the children of other campers came over to meet our boys and invited them to join the gang. We never met other American campers and only occasionally Western Europeans. In one campground, a young teenage girl in a white shirt and blue cotton skirt came over with a couple of yellow plastic pails. Then she realized our boys didn't speak Russian. She took them by the hand, gave them each a pail, and they joined a group picking raspberries in the woods. In all cases, the older children took initiative and responsibility for the younger ones while the adults did their adult things.

Each campground had a camp store. Campers would enter a small building with a long counter in front. At one end sat a big glass bowl filled with water in which a hunk of butter was floating to keep it from melting. Partway along the counter, a large bottle of vodka and one small glass were usually at the front of a line of campers waiting for drinks. The first in line filled the glass with vodka and drank it down right on the spot. The glass was then passed to the next in line, or maybe the first had a second drink before passing the glass. A cashier behind the counter collected the cash for the vodka and typical groceries such as small tins of pineapple from Cuba (a Communist ally), tan paper sacks with noodles or flour, and thick slices of heavy brown bread sold by weight.

In one campground, Hajo and I were invited to an evening campfire where campers were roasting potatoes in the coals. We all sat on the ground around the firepit, singing traditional Russian songs as the potatoes cooked. Then we helped dig the potatoes out and passed them around the circle. Soon, everyone began eating the potatoes, burning fingers and lips on the hot skins and cooling their mouths with potato vodka!

Vodka seemed to be the drink of choice wherever we were in Russia, but kvass was also available on many street corners. We'd see an old wagon carrying a big metal barrel on the back with a spigot at the rear. A person could bring their own container to fill to take home, or they could drink it on the corner. I tried it one time, turning the spigot and filling my mug with a brown liquid that resembled a dark beer. But it was kvass: water with slabs of old, dried brown bread soaked in the barrel until the whole began to ferment. Then it was ready to drink. I can't really describe how it tasted—not sweet, not sour, just bland—but one drink was enough. When I returned to Russia in 2018, I saw no sign of kvass, and when I asked young people about it, they didn't know what it was.

Our VW camper was a sensation in Russia. Most Russians waited years to purchase a basic small car. One campground was built on a series of three terraces, which sidestepped down a slope to a good-sized ditch with water slowly flowing along. The driver of one of these small cars was apparently tired as he arrived at the campground. He parked his car at the top of the terraces but forgot to set the brakes as he visited the nearby camp store. We watched from afar as his car rolled down the several embankments and ended up half-submerged in the ditch. Immediately, half a dozen campers, who had seen this happen, grabbed ropes and ran down to the ditch. They connected the ropes to all possible places to keep the car from sinking further or floating away. Some men stripped down to their shorts, slid into the ditch, and fastened more ropes to the fenders and wheel axles. Then about ten men pulled and pushed the car up from the ditch and up the embankment as water gushed from the trunk, doors, and hood. The owner came back to see his rescued car already being cleaned up and prepared for his return. This was socialist cooperation at its best! During the next few days, the mechanics among the campers helped him get the car in running order. Gratitude was written all over his face!

How about a Camping Trip in the Soviet Union? | 93

Meanwhile, other campers approached and timidly asked us to show them our camper. They had never seen camper vehicles before and were amazed at the beds, stove, sink, and storage space, as well as the seats for travelling. We also brought a Polaroid Instamatic camera. We asked if we could photograph one family. They posed, and a few minutes later we gave them the Instamatic photo as a souvenir. They thanked us with four small green cucumbers, some of the rare fresh vegetables we enjoyed on this trip.

In every campground, we were required to register with an official host who was obliged to stay on duty until everyone on the official list of international campers had signed in. I remember that in one campground, our young, hospitable hostess spoke excellent French. She checked us in, then added that there was only one camper yet to arrive before she would be off duty. Later in the evening, I dropped by her small office to ask her something and discovered that the delinquent camper had not yet arrived. The hostess seemed bored, and happy to have someone with whom to speak French. We hit it off, as they say!

I asked about her family. Her older brother was in the Russian army in Afghanistan. She was proud of him for "bringing civilization and order to this backward, poverty-stricken, violent land." I hope he made it back home to her. I wish the US had learned from the Russian failure in Afghanistan before our own engagement and failures. Meanwhile, the expected camper did not arrive, and the local militia was contacted to retrace his route and paper trail until he was found.

I learned firsthand the element of fear used in totalitarian societies. We had been told not to photograph airports and I didn't. But somewhere west of Moscow I saw a small biplane, a true relic, flying right overhead, and I snapped a photo. A few minutes later, the police stopped us. It wasn't the usual routine check. Had I taken a forbidden photo? I didn't know and nothing was said. An hour later, we arrived in the outskirts of Moscow at our assigned hotel. The reception officials said they did not have two adjoining rooms. Hajo and Eric would have to stay in a room several floors up from Mark's and my room at the opposite end of the hotel. I was still shaking from having taken the biplane photo. I felt our rooms were probably bugged, so we didn't discuss anything of substance. A woman tending a samovar sat on the end of each corridor, watching the comings and goings of the people on her floor. I wasn't sure what her role was. I think she mostly ignored me when she found out I didn't speak Russian.

I remember how ill at ease I felt not knowing what was allowed and what was forbidden. We had been required to pay for our meals at our hotel, but had not been told we needed reservations, so we waited for two hours to be served a cold, tough piece of some meat and a large bowl of hard, old green peas. I remember the peas because they were served again the next morning for breakfast. Meanwhile, our servers just stood around conversing in a corner and ignoring the customers, presumably because they were paid a set salary whether or not they waited on people. Between fear and a lack of service, we were not thrilled by official Russian hospitality!

In Moscow itself, we were guided to the usual tourist sites, including Red Square and the many-turreted Saint Basil's Cathedral. We stood in line at Lenin's mausoleum to see him lying in silence as many Russians, obviously moved, filed by his glass-covered coffin. We attended the circus, where we saw astonishing acrobats and clowns. At a symphony concert, I managed to translate a program printed in Russian from the Cyrillic alphabet to know we were hearing Beethoven and Mozart. We were visiting a memorial to the "heroic" Russian victims of World War II when we heard groups of schoolgirls speaking British English. They told us they were from an English-language Russian school. I wanted to know more about it, but their teacher caught them talking with "the enemy" and quickly hurried them away from us.

Later in Kiev, we had a very knowledgeable Russian Intourist guide, Ivan. He was happy to be assigned to Kiev and show tourists around the city. We invited him for supper at our campground. After the meal we sat at the picnic table in the campground, drinking vodka. I had definitely had more than my usual amount and lost my inhibitions. Ivan was saying he had become tired of tourists insisting their rooms were bugged or that they were being spied on. I said I did not usually feel that way, but wondered if a man leaning against a nearby tree was listening to our conversation. Ivan's face fell as he turned slowly to look. There was no one.

This was not an issue taken lightly. Ivan thanked us for the supper and left. Another official guide was assigned to us for our last day in Kiev. Part of this day I spent with our boys, playing in the sand by the Dnieper River with two young Ukrainian boys and their mother. It was a lovely, sunny, idyllic morning. Now, in 2024, as the Russians continue their attacks on Ukraine, I wonder: what has happened to this mother and her sons so many years later?

I could write more about magnificent cathedrals, icons, gold domes, palaces, and places we visited. I remember that after three weeks, as we waited at the border to cross into Poland, we had had enough of the Soviet Union. Russian guards took apart everything in our camper. They went through my purse and paged through my address book. They looked through our dirty clothes bag, our camping gear, the engine compartment, and the undercarriage of our camper with long mirrors. They found nothing incriminating and took back all our official papers except our passports, which they returned.

We crossed over the patrolled border. The Polish guards greeted us warmly. We asked if they needed to examine our camper. They said, "Not necessary. Our brothers across the border did a thorough check. We watched them! You are free to go." And they laughed! We breathed a sigh of relief as we drove west.

One Summer in Israel: "L'Chaim—To Life!"

In 1980, I was fortunate to receive a scholarship to spend the summer in Israel doing anything I wanted, provided I agreed that when I returned home to Havre, I would integrate my Israeli experiences into my teaching and talks in the community. I experienced Israel from many different perspectives, and since then, I have shared my experiences not only in Havre, but around the world. Here are a few of the events and relationships that made the summer so special.

Before my stay in Israel, I had read James Michener's *The Source*. Michener describes the journey of a young woman who goes early one morning to fill her pitcher in the deep well of a Galilean city. To reach the water, she must descend many winding steps to the bottom of a tell, the mound of successive layers of civilization built up over the centuries. As she walks downward, she rouses two large, intensely blue birds, which burst upward, circling from the well pit. One of my first mornings in northern Israel, near dawn, I descended into the ancient well of Hazor on the plain below Mount Hermon. As I walked downward into the coolness, I was startled by the unexpected explosion of beating wings and the flight of two incandescent azure birds, which soared upward above me and disappeared over the old walls. Inspiration for Michener's book? Inspiration for my day!

The Sinai Desert

During my stay in Israel, the security situation was variable. I could hear firing on the Golan Heights. The Sinai Desert was still divided between Israel and Egypt. However, it was open for a ten-day expedition in a six-wheel-drive, open truck-type vehicle. I was able to get a reservation and had one of the most fascinating trips of my life! One of our guides had just completed his military service, which included guarding the border between Egyptian and Israeli outposts in the desert. He knew the desert, from every wadi (dry riverbeds that flood when it rains) to the numerous oases. I did not realize how very many oases there were, all formed by eons of geological developments in the Sinai. I had imagined Moses leading the Chosen People through a waterless wasteland, but we visited oases and wells of many kinds every day.

Our guide explained that the military outposts in the Sinai were so isolated that the Israeli and Egyptian soldiers guarding the borders held parties together to break up the boredom of their terms of service.

While touring along a rough wadi, we suddenly came upon a lone tree, the only one we had seen for hours. The driver stopped and we saw a small, carefully scripted sign: Sinai National Forest!

Among the participants on this expedition was a university student from an Orthodox Jewish family in Jerusalem. She was trying to make major decisions in her life. Her Orthodox parents were demanding that she abandon her studies, marry an Orthodox man, and begin bearing Orthodox children to populate Israel. They were so insistent that they threatened to disown her if she didn't give in to their wishes. I have always wondered what she decided and what the repercussions were.

Another participant was a young woman hired by Club Med to accompany one of their members who chose to go on this expedition. She was a French speaker, and I enjoyed talking with her when her charge was relaxing and reading. After our first day together, she said she wanted to tell me about her background, but I wasn't to talk with anyone else on the expedition about it. I agreed, and she trusted me.

She said she was of Romani—or tzigane—background, and had spent much of her life as a kind of modern nomad wandering about Europe, particularly in France with her birth family. She didn't want people to know her background because she said people would think she was a thief. She explained that in her tradition, it was acceptable to take things if someone really needed something. Suppose a person found something really useful that was sitting around not being used—a person wanting it could take it with good conscience as a gift from God.

She also said that she was a few months pregnant. As soon as she started showing, she would lose her job with Club Med because she wasn't married. She said she got pregnant on purpose because her grandmother had told her a long story of her heritage and family going back many generations. The story took three years to repeat, and her grandmother had told her the story three times. She was the carrier of the story and had to tell it to her children to keep it alive.

I asked if she couldn't write it down. She explained it was part of the spoken tradition of her family and had to be kept alive orally. She had not yet met a man she wanted to marry. But she had chosen an intelligent, healthy man to father her child, whom she hoped would keep alive the family story. I asked her if we might stay in contact after the expedition. She said she didn't have a fixed address and that we would meet again somewhere, sometime, if God willed. Thus far, we have not yet met again. But I am happy to have had contact with a person of her background, so different from my own, and to have heard her side of the story.

We visited many fascinating sites in the Sinai, including Mount Sinai, which is regarded as the traditional mountain where Moses received the Ten Commandments. Monks from the monastery at the foot of the mountain had cut hundreds of steps up the mountain. We climbed them in the dark, reaching the summit just as the sun rose over the surrounding desert peaks. We felt we had entered a sacred space as we sat silently in the warming sunlight. It seemed natural to think that Moses had transcribed the Commandments on stone because that was the only material available.

To me, much of the Sinai had a mystic or spiritual feel to it. One morning, several of us rose early to go for a swim in an immense natural rock cistern some one hundred feet long and deeply filled with water. The cistern was in deep shadows except for a shaft of light that shone through a gap in a rock wall. I was drawn to this light as I swam to the far end of the pool.

Here, I observed a silent miracle of rebirth. Several small rocky concretions caught my eye. They looked like capsules of tiny pebbles glued together. But what held my interest was the emerging forms of the glistening silver bodies and wings of what looked like dragonflies. They were struggling to free themselves from the confining forms of their encrusted shells. As they emerged from their chrysalides to their flying form, I realized that they, like us, are meant to be free. This was an Easter metaphor right before my eyes. The power of resurrection helps us move beyond the confining shells of prejudice, fear, selfishness, and bad habits if we give it half a chance.

One day we rested at a desert well where the water flowed so plentifully that a small oasis had grown up around it. Here in the shadow of a vine-covered trellis, I looked out across the rocky desert floor to the rugged cliffs of a red canyon wall. Everything was silent in the heat, but then I heard a high, haunting melody. I could not discover its source. Then I saw a distant camel moving toward us. Perched up high on its back sat a young desert lad, his simple cloak billowing in the hot wind, his voice flowing out in clear joy! He came to the well and greeted us openly as his camel knelt and he descended. We talked a while as he drank contentedly of the cool water. Then he took a length of hose that ran from the well, primed it, and watered a few pepper plants and some tomatoes and melons in his father's garden. As simply as he had come, he mounted the camel and rode away again, singing. Soon, the echo of his melody was all that was left, along with my own desire that I might live and work with such grace.

Our expedition took us to a broad beach on the east coast of the Sinai Peninsula. It was covered with hundreds of beautiful conch shells, as well as thousands of tattered plastic bags that had been blown across the Gulf of Aqaba from various Middle Eastern countries. Such a contrast of delightful and disgusting. That was typical of our desert expedition. A Bedouin woman sat with us one day, showing us how to apply makeup in a desert way. I won't say what we looked like! On another day, most of us came down with amoebic dysentery from drinking from the desert wells. It took us weeks to heal! But now, years later, I mostly remember the inspiring times: snorkeling in the Red Sea with multicolored fish, seeing cairns of rocks marking trails through the desert, and more than I can list here.

Jerusalem

My month in and around Jerusalem was filled with classes at Hebrew University, archaeological digs at an ancient dump in the City of David, trips to Jericho, and concerts and tours of various kinds. The concert I remember most was Itzhak Perlman (yes, the same man Pop spotted fly-fishing near the Maroon Bells in Colorado) playing a welfare concert for physically handicapped children in Israel. We saw the world-renowned Jewish violinist slowly crossing the stage on his crutches to greet several dozen children with mobility problems, who were seated on the stage waiting for him. Then he sat with them and proceeded to play exquisite pieces as we in the audience looked on during this emotionally charged evening. The power of genuine caring and spectacular music was unforgettable.

Most of my time in Jerusalem was spent at the ancient City of David dump, where we sorted through pottery shards from Roman times, looking for anything indicating that the Israelites had lived here before anyone else: a way of proving they had rights to this land before any other peoples. I found lots of pitcher handles and oil-lantern pieces, which were thrown on the "destroy" or "souvenir" piles. So far as I know we found nothing of great archaeological value, but I loved the digging and looking. And I enjoyed working with the enthusiastic Jewish students and immigrants. They were dedicated to leading a peaceful, creative, normal life where they were accepted and could be proud of their identity.

It was inspiring to work across the valley from the Mount of Olives, which was the scene of many significant events in the life of Jesus. We also took a field trip to the Pool of Siloam, which is mentioned in the Gospel of John. Places like these meant more to me than the churches and cathedrals built on holy sites, which destroyed the original feel of natural locations. Being in the Holy Land seemed to make the Biblical scriptures jump from the page; the "stones cried out" in so many passages, and the land was littered with stones. For example, the Parable of the Good Samaritan begins, "A certain man was going down from Jerusalem to Jericho." I was on the road that twisted down over three thousand vertical feet from Jerusalem to Jericho over a very short distance. There were multiple opportunities for an ambush. Little wonder the man was attacked and everyone passed by on the other side. The Samaritan was putting his own life on the line to stop and help him in such circumstances. Being on this road myself gave me a powerful message about helping one's neighbor even in the toughest of situations, even if he was a foreigner.

    I was moved by the visits to the Wailing Wall, where practicing Jews prayed and left paper prayers in the cracks between the large remaining stones. The long history of Jewish suffering up to the present is so powerfully expressed in this holy place. And nearby, the Temple Mount, with its splendid architecture, calls Muslims to worship and tourists to wonder at the beauty of the site.

Family Adventures

The last part of the summer, my husband and two sons came to Israel, and we toured together. We stayed in a campground run by Palestinians on the outskirts of Jerusalem. We rented two tents for several days. One day, we left early in the morning for a drive up to Galilee. About noon, both boys started feeling ill with headaches and stomach issues. With their ailments, it was too far to drive back to the campground, so we stayed in a hotel.

By the next day, they were well again, and we drove back to our campground. When we arrived, we saw that our tents were gone! We hurried to the campground office to ask what was happening. The Palestinian manager apologized profusely. The Israeli army declared they needed the campground immediately for a gathering of young Israelis who had their military service meeting, and the Palestinians were obliged to clear the campground for the Israelis, no exceptions, not even if it meant getting rid of the other campers. The Palestinian manager indicated another campground where we could go for the rest of the week. He had carefully collected all our belongings from the tents, including a wallet with money and a watch, which he gave back to us. He showed the utmost honesty and concern as he helped us load the car and move on. I was really impressed. The Palestinians could have taken advantage of our absence, but they didn't!

Meanwhile, we stopped by a small grocery store to buy a few supplies. As I was getting out of the car, I saw an Orthodox man approaching on the sidewalk. It was very hot, but I knew I had to cover my arms to respect the Orthodox traditions. I quickly put on my sweater, but not fast enough. He grumbled something under his breath and clapped his hands over his eyes to avoid seeing my sexy elbows! Then my husband and boys entered the store, with me at a distance behind them where a woman belonged. In the store, a woman and a man were in a heated argument. The next thing I knew, we were in the middle of a barrage of yogurt as the man threw several containers of it at the woman, who ducked away behind a series of shelves. We decided we would do our shopping elsewhere as I followed behind my menfolk back to the car.

We visited some German Jewish friends of Hajo's parents. They had survived the Holocaust and immediately immigrated to Israel. They had planned to visit Europe the summer we were in Israel, but we discovered they were still at home in Jerusalem. They explained, "We're staying here. If the Palestine forces on the Golan Heights see too many Jews leaving Israel, even on vacation, they will think they are winning!"

We camped by the Dead Sea, much as we had done by the Great Salt Lake, but Israel was incredibly hot, even at night! We lay on the ground with wet sheets over us to be cool enough to sleep. We visited Masada with its brutal stories of Jews and Romans at war and its scores of interpretations. The site itself is remarkable, situated above the Dead Sea with its cliffs and ramp. We also made desert hikes up valleys watered by streams and waterfalls.

The summer in Israel was over all too quickly. It left me aware of many conflicting issues. I appreciated that Jews of various backgrounds and theological positions needed some country to call home, where they could feel that they belonged. But I could also understand that the Palestinians had lived in the area for centuries and considered it their home as well. Many Jews said that Palestinians could easily move to other Arab areas where there was land enough for them. There were attacks from both sides, trying to protect their homes. Meanwhile, now as then, there are groups of both Jews and Palestinians trying to work together to solve major issues peaceably, even as the Israeli military and various Palestinian militant groups kill each other in tragically destructive, violent conflicts.

Slip Sliding Away

Hiking is what my family did during the summer. Skiing is what we did in the winter, at least when it was possible. I started skiing before it was big business, before there were major ski areas in Colorado. When I was ten years old and there was enough snow, I'd take an old pair of child-size skis up to Chautauqua Hill under the Boulder Flatirons. Sometimes there was a rope tow rigged to pull skiers up the hill. Usually, I'd walk up the hill until I was tired. I'd put on the skis and snowplow down the ungroomed slope until I fell. Then, I'd pick myself up and do it all again. I liked the sensation of sliding downhill!

One snowy Saturday, I heard that someone was sponsoring a ski-jumping contest at Chautauqua. All we had to do was show up. I was in the children's division. There were four of us. Our jump was really just a mogul about two feet high. Some older children went down the steep entry slope and over our jump to show us how it was done. They made it look easy. None of us had ever jumped before.

The first contestant headed down the hill so slowly that he almost stopped before the jump, but he made it over and managed not to fall, so he wasn't disqualified. The second jumper sped down the hill, went sailing over the jump, and landed on his back, disqualified! The third contestant slid down the slope primed to jump, flew through the air, and landed on his skis without falling down. He looked like a pro! I was last. I knew I could not jump as well as he had. I started down the hill and picked up speed that took me over the jump, but not as far as the third guy. And I remained on my skis without falling! I was in second place!

Then came the awards. First place was a shiny gold cup which had "1st" engraved on it. My competitor held it up proudly! I was waiting for my prize. The announcer said, "Second place: a pair of skis!" I couldn't believe my ears. I received a real pair of skis with metal edges and metal-and-leather binding devices. I was thrilled. I had my own pair of real skis. I used them for ages until I moved to Montana and received a new pair when I was accepted on the ski patrol.

My skiing was mediocre compared to the other members of my family. My brother was a cross-country skier with the Colorado University ski team. Later, he did his military service as a biathlon army team member. He also ran the cross-country ski center at Steamboat Springs, where Pop also led ski tours. My nephew Peter, son of my sister Sue, skied as a member of the US Olympic Cross-Country Ski team and later coached the team!

While they were competing, I was skiing for fun.

When Hajo and I lived in Germany, Elisabeth, our exchange student from Norway, invited me to go ski touring with her and her friends on the tundra trails from hut to hut during their Easter break. There was plenty of snow up high. We could look down over the cliffs and see the fjords below with their spring-green meadows and dozens of Norwegian flags flying for the Easter festival. It was such fun to ski across the miles from one hut to the next. They were all run on the honor system. Our overnight rates and food costs were listed on a board, and everyone paid accordingly. No one ever thought of cheating.

I loved to see the Norwegian skiers making smooth telemark turns down the mountain slopes. Such skill and grace! We'd ski all day and arrive at a hut near evening. I remember that at one hut we were all frantic. We could not find the usual candles and lanterns. There were matches for the stove, but nothing for light. Then someone saw a switch on the wall. He flipped it and we had light! This was the first hut to have solar panels. No one had bothered to tell us, and we hadn't noticed them when we arrived!

As often as I could during our twenty years in Montana, I either went cross-country skiing or skied with the ski patrol at several small areas that were just developing. One of our duties was to check skiers for frostbite and convince them to wrap up! Some days were so cold that I had to push with my poles to ski downhill. When the snow gets too cold, there is simply too much friction to ski. When it's that cold, most people stay inside!

When we returned to live in Colorado in the 1990s, I made many ski hut trips. One memorable outing was with Anne Gini, a Norwegian former cross-country ski instructor. Frances, Margo, and I were three of her students. One winter, we rented a yurt for a few nights in North Park near the Wyoming border. We drove on plowed back roads to a parking area where we left the car and took off skiing to the yurt. Frances pulled a special sled with food and supplies so we could party! As the sun went down, the temperature fell accordingly. We got a fire going in the yurt stove, and the place warmed up amazingly given the outside temperature. We splurged on spaghetti with meat sauce and trimmings and red wine.

Unfortunately, in the middle of the night, we had to trek out to the toilet a few hundred feet away from the yurt. Placing the toilet away from a hut or yurt is done on purpose, so that if the hut were to catch fire, the skiers could survive in the outhouse! A good thought for sure, but that night we wished the toilet was closer in. A thermometer on a pole near the yurt registered thirty degrees below zero! We survived, but a day later when we skied back to the car to drive home, it wouldn't start. There we were, stranded miles from the nearest ranch. Fortunately, after we waited for the sun to shine on the car, it warmed up enough to start, and we made it back to civilization.

Eldora Ski Area in the mountains above Boulder includes groomed cross-country trails with wilderness on all sides. I loved to go up there and glide in the set tracks. One day in around 2015 I was skiing there when I came around a downhill curve very fast. I knew the course and that I could make the turn without any problem. However, as I was half-flying around the track, I suddenly encountered two huge female moose, seemingly gossiping together. They took up the whole trail! I was going too fast to stop. If I sat down, I would careen under their legs. If I cut off into the forest, I would kill myself on a trunk. As the options raced through my mind, the moose instantaneously split. One ducked into the forest to my right, and the other to my left. I sped right down the track where they had been. Whew!

After World War II, the 10th Mountain Division created a system of huts in the Colorado mountains. Some were within a few miles of a road, others ten or more miles in. It was always a challenge to ski to them because my backcountry ski group needed to carry not only the usual sleeping bag and food, but also survival gear in case we were caught out and needed to bivouac or to dig someone out of an avalanche. One trip, we were to share a hut with another group. While we were waiting for them, we heard the motor of a small plane overhead. It circled the hut and then dropped a large bag with a note. It contained a complete turkey dinner with trimmings, a gallon of ice cream, and cake. They had solved the weight problem!

Backcountry skiing is a grand adventure, gliding slowly through the snow into the wilderness. To ski unbroken slopes is what every skier dreams of. The possibility of avalanches is always an issue. One of my friends, who taught in an avalanche safety school, was herself taken down by an avalanche despite her knowledge. Fortunately, the people she was with were able to dig her out before it was too late. She was humbled by that experience. I leave scary backcountry skiing to others.

Figure 17. Floating down the Missouri River

Figure 18. Hiking in Glacier National Park

Figure 19. Typical campground in Russia

Slip Sliding Away | 111

Figure 20. I love cross-country skiing!

Figure 21. Foreign exchange "sisters" Elisabeth and Freddie meet for the first time

How I Moved beyond a "Whites Only" World

While I was teaching school, helping raise two boys, digging dinosaurs, watching the eclipse, and floating down the Missouri in rural Montana, the Civil Rights Movement was sweeping across America. I followed it on television as I remembered my own experiences of racial awareness and exposure. I lived as the terms changed: Negro, African American, Black, Black and Brown. It was hard to know the vocabulary of the moment. I will use the terms that seem appropriate to the time frame for each story. For example, I use the word "Negro" in the context of the Civil Rights Movement to reflect the terminology preferred by many Civil Rights leaders at the time, while changing to "Black" in other sections to reflect modern preferences.

When I was about five years old, I lived in Boulder. My next-door neighbor was Marshall, a boy from Australia. I don't know where we got the idea, but we invented our own game, "Mama Drink Tree." In his yard, there was a large apple tree with limbs where we could climb and sit. Each limb represented a "mama" of a different race whose breasts produced drinks of the appropriate color and flavor. We moved from limb to limb, pretending to suck the drink we desired. I wonder if this early game fostered my lifelong interest in the diversity of peoples and cultures.

Exposure in School

For me, the Civil Rights Movement was a distant reality. I was totally unaware of the small Negro community in an area of Boulder. But I learned about Gandhi in elementary school and the injustices of colonialism. I remember when two teachers from Germany arrived at our school shortly after the end of World War II to learn about education in a democracy. In junior high, discussions about starvation in third-world countries caused enough concern to keep me awake at night. We read a *Weekly Reader* article about segregated drinking fountains in the South and Negroes having to ride in the back of the buses. These were the secondhand incidents I was about to face in person.

A Trip through the Deep South

In 1952, when I was fourteen, my family drove during our Christmas vacation from Boulder to Miami Beach, Florida, to visit our Aunt Mary. This trip took us through the Deep South before there were interstate highways. The life we saw along the roads opened my eyes to the realities of segregation. I remember the dilapidated grey clapboard houses with rickety steps leading up to crooked porches, in front of which dozens of barefoot Negro children were running and playing in dirt yards. We also drove past long, shaded driveways with grand mansions at the end, which we glimpsed through Spanish moss hanging from huge old trees. I had a vague feeling these were part of a slave period that was long ago. In the towns, I saw "Whites only" signs. I remember thinking, "I'm seeing what we read about," but I had big gaps in my knowledge of Negro history and life in the South.

En route to Florida, we stopped in a Mississippi town in front of a drugstore. My dad said to me, "Pat, hop out of the car and get me a box of Sucrets. My throat is bothering me." I left the car with some loose change in my hand and walked into the drugstore. A bell jingled as I entered. To my right, a half dozen Negro folks seated at a table with coffee cups simultaneously stood as I walked in.

To my left I saw a glass-fronted counter. I timidly walked over and asked a Negro man in a white coat who looked like a pharmacist for some Sucrets as I placed the coins on the counter. I remember him saying, "I'm sorry. We don't carry that brand of cough drops here, but if you go a few blocks down the street you can probably find some in the white drugstore." I thanked him, picked up my change, hurried out the jingling door, and heard chairs being pushed toward the table and folks sitting down. I remember thinking there was something terribly wrong when Negro adults felt compelled to rise in the presence of a white child.

Native American Adventures

During my high school days, I was active in the youth group of the local First Baptist Church, part of a very liberal, social justice–oriented denomination. Every spring vacation, we made trips to the Navajo and Hopi reservations in Arizona or the Oklahoma Native American areas for various work projects. During one trip in Oklahoma, we used a local bus for transportation. We Boulderites took seats about halfway back in the bus. At one stop, a dozen Negro students boarded the bus. They indicated we were sitting in their seats and should move to the front, so we did. A few stops later, a group of Native American teenagers entered the bus and immediately moved to the far back. Maybe I was wrong, but it seemed to me that the racial pecking order was well established on the city bus.

On the Navajo and Hopi reservations, I experienced firsthand the issues of reservation life. The Navajo families lived in traditional hogans or government houses in isolated areas on the reservation. In those days, parents hid their children so they would not be forced to attend boarding schools far from their families. Many of the residents did not speak or understand English, and government translators were hired to communicate between the Bureau of Indian Affairs and the Navajo families. I feel ashamed now that, at first, I thought their religion was superstitious and inferior to Christianity. I have since realized the value of visiting the reservations and coming to know firsthand the people, and their ways of living and practicing their beliefs.

I feel particularly privileged to have been invited into Hopi homes. In one, the mother was making traditional rolled paper bread from blue cornmeal. She had an open fire on which there was a large flat stone as hot as a grill. In a pan, she had mixed blue cornmeal with water. She dipped her hands in this mixture, which she spread in a thin layer on the stone. It was so hot that the cornmeal immediately cooked and began curling up into a roll like a blue newspaper. She stacked these rolls against the wall for her family to eat whenever they were hungry. Now, some sixty-five years later, American shoppers buy blue corn chips at the supermarket.

Florence Means

During my time in Boulder, I had numerous contacts with the author Mrs. Florence Means. She wrote dozens of books for children and teenagers about American minorities, including Negroes and Native Americans. In 1945 she wrote *The Moved-Outers* about a Japanese girl in one of the internment camps during World War II. For this book, she received a Newbery Medal. She showed special courage to confront this injustice immediately after it happened. People like her influenced my broader vision and concern for minorities.

Chikako Ando

Mrs. Means was not the only one who had firsthand experiences with a Japanese girl; so did my family. In addition to his work at the University of Colorado, Pop worked for a government agency that was responsible for the accreditation of high schools. In this role, the US government sent him to inspect American schools for the children of military personnel in Japan. After a family discussion, we decided in 1956 to invite a Japanese girl to live with us as an exchange student.

While Pop was in Japan, he interviewed many girls interested in spending a year in America. He chose Chikako Ando, an outstanding student who later became a physician and the head of a Red Cross Hospital in Japan. Chikako became a part of our family and the school community. She seemed to be accepted everywhere, and as far as I know there was no issue of discrimination. She went camping with our family, attended school dances (which were completely unknown in Japanese high schools), and joined the school chorus. Mrs. Means invited Chikako and me to tea several times during the year, which we both enjoyed. We remained in contact with Chikako for over fifty years.

Summer Exposure in NYC

Mrs. Means also contributed to a fund that enabled me to volunteer in New York City, working with inner-city children and teenagers, the summer after my sophomore year in college. I lived in student housing on Washington Square and worked through Mariners' Temple Baptist Church on the Bowery. My job was to give inner-city Negro and Puerto Rican children experiences beyond the high-rise building projects where they lived. I rode up elevators that smelled of urine or trudged up dank stairwells to meet the families of my young charges. I met Daniel who, as a toddler, had fallen five stories down an elevator shaft and seemed unscathed from the fall. His white mom had married her Negro husband in Belgium when he was stationed in Europe with the American army. She was having a hard time adjusting to minority American life in the inner city and was relieved to speak her native French with me. There were also the Jones twins, girls who, like many of the children, came on our field trips with potato sandwiches: two slices of white bread with mashed potatoes spread between them. I doubted the children were getting the nutrition they needed.

One of my responsibilities involved coaching a girls' softball team. Two groups of teenage girls appeared for practice, the Puerto Ricans and the Negroes. They disliked each other from the start. The father of one of the Negro girls was helping to coach this team, such as it was. He had been a minor league baseball player and knew his ball. He told both groups of girls, in no uncertain terms, "to get their asses out there and play like a team!" The girls did the best they could, though they had never played ball before. We didn't win a single game, but the girls had a chance to get out of the city and play teams in New Jersey and upstate New York. They said they were glad that they lived in Manhattan because life was so "boring" everywhere else!

One Sunday that summer, I was invited to go with members of a Negro congregation to a church picnic on Long Island. It wasn't like any church picnic I had ever attended. We all arrived in dozens of buses from all over the New York City area. We mingled around long wooden picnic tables, eating our fill, until someone started playing a saxophone. A clarinet picked up the jazzy tune, and then guitars and drums joined in. We were all up on our feet, dancing around the tables and down the roads, clapping and singing. It was so spontaneous, so fun! And so moving too, when the traditional gospel songs echoed around the picnic site. All too soon, we had to climb on our buses and ride back to the city. I remember returning to my student house and washing up to go to bed. The white face looking back at me from the mirror was the only one I'd seen all day.

Meeting Martin Luther King Jr.

This was the same summer that Martin Luther King Jr. addressed a group of us summer interns. He said that the Civil Rights Movement wasn't only to bring equal rights to Negroes, but diversity to white people to enrich their lives. We didn't know what we were missing in our whites-only world! That was a new idea to me, and I have never forgotten it.

After his presentation, he did not disappear backstage as many speakers do. He came right down into the audience, walked up to me and the group I was with, extended his hand, and said, "Hello. My name is Martin. What is yours? What can we do to help you with your work here in New York?" I was astounded. A person of his caliber coming up to me, taking the initiative to introduce himself and show a personal interest in our work. He was "walking the talk," as we said in those days. This short contact changed my life. I realized it was up to me to take the initiative to meet people, to make things work.

University Experiences

I returned to the University of Colorado and joined the student chapter of the NAACP. This was in the mid-1950s when the Civil Rights Movement was picking up steam. We decided as a chapter to join the picket line at our local Woolworth store. Even though the Boulder Woolworth on the corner of Broadway and Pearl did not have a lunch counter, we picketed to change Woolworth's national policy. It was unconscionable that Negroes were not permitted to sit down for lunch and had to stand at the far end of a lunch counter. We demonstrated to change this practice. Eventually, Woolworth did change its national policy. I remember one of my mother's friends seeing me with a picket sign. She looked straight through me, pretending not to know me, and walked straight through the picket line and into the store. I felt like we had a long way to go for equality to be a reality.

About the same time, the University of Colorado hired its first Negro professor, Dr. Charles Nilon. He was a member of the English department. I was fortunate to be in one of his first classes. I remember how he brought poetry alive for us.

He and his wife attended the liberal First Baptist Church in Boulder, where I was also a member. The pastor knew the Nilons were having a hard time finding an appropriate home in Boulder, and he asked some of the realtors in the congregation if they could help. During this era, it was said that Negroes moving into a white neighborhood would cause house values to fall. Shortly after the pastor's request, one real estate agent left the church. The Nilons finally obtained a home near the traditional Negro neighborhood in Boulder. (It would be twenty years later before Bill and Eileen Klein, Negro teachers at Boulder High and CU, purchased a home "on the hill" near the campus and next to my family home. None of the house values dropped, and they were accepted as good neighbors.)

Dr. Nilon also invited interested students to his home on Saturday evenings to listen to his extensive jazz record collection. I had had very little exposure to jazz before and felt out of my element. I owe Dr. Nilon for this valuable introduction to various jazz styles and performers.

I was busy that year. I joined an interdenominational youth group and was selected as a representative to a national conference in Wooster, Ohio. My conference roommate was the daughter of the Negro dean of Tuskegee Institute. We walked over to a restaurant near the campus for a hamburger. We sat in a booth chatting as people around us received their orders. The waiter ignored us.

"You know why we aren't being served," my companion said. "We might as well leave." The reality hit me in the pit of my stomach. I felt sick. I couldn't believe what was happening. Ohio was in the North. This wasn't supposed to happen in the North. I remember thinking, "Those hamburger flippers have no right to refuse service to the daughter of a university dean just because her skin is dark." I was beginning to learn that prejudice has no borders.

Prejudice in Montana

So much has been said about segregation in the South, but even in Havre in the far north of Montana, I was shocked to see how blatant prejudice could be. For example, when Martin Luther King Jr. was assassinated in 1968, one of the town's businessmen declared at a Kiwanis meeting, "Well, they finally got him!" I wished I had been at the meeting to express my reactions.

Buffalo Soldiers and Beyond

It was general knowledge in Havre that General Pershing had led Negro infantry troops in Montana after the Civil War. They were stationed at various forts in the state, including Fort Assiniboine just west of Havre near the Milk River. They were part of the Buffalo Soldiers units formed to protect white settlers in the West from hostile "Indians." This unit of infantry soldiers rounded up bands of Cree still in Montana and deported them to Canada. In 1916, the US government invited the Cree and related Chippewa people back to Rocky Boy Reservation in the Bears Paw Mountains southwest of Havre. The local Chippewa Cree dislike the name "Rocky Boy" because it is a poor translation of "Chief Stone Child," a highly respected chief for whom the reservation was named. After about ten years, the Buffalo Soldiers left Montana to fight in the Spanish–American War.

There are still complicated issues between local residents and the reservation tribes in Montana. For example, children from Rocky Boy Reservation are eligible to attend either a reservation school, a small-town school in Box Elder near the reservation, or one of Havre's public schools. Some of the children begin the school year on the reservation, but then decide they would rather go to school in Box Elder with some of their friends, or in Havre to play basketball. The switching of children from one school to another causes difficulties, as it is impossible to know whether they are skipping school or attending somewhere else. To deal with this and other issues, Havre Public Schools has appointed a Native American education coordinator who works to improve the situation.

There were several different Native American tribes in the Havre area. When I was teaching there, approximately 5 percent of the local public school and local college population were Chippewa Cree or other tribal members. These numbers have increased since then. I was the "Indian Club" sponsor at the Havre junior high school for a number of years. The club sponsored activities and trips for tribal children and their families. Some of the girls wanted to learn to bead, so one of the elder beaders came to Havre and worked with half a dozen girls one afternoon. All the girls worked at it for about an hour and then decided it was harder than they thought. The elder showed great patience and explained how to finish their project at home. She volunteered to come again, but they thought it was too tedious. That ended that project. (However, an adult education class in beading was later offered.)

About the same time, the Indian Club sponsored a bus trip to the Blackfoot Reservation pencil factory and museum in Browning, Montana. I remember waiting with the students and their families for the bus to arrive when two very large, dark men approached me. One was on my right side and the other on the left when they both started squeezing me in the middle. "What are you doing?" I asked, a bit intimidated. "

Oh," one said, "We are playing Oreo cookie. White icing between two chocolate cakes. We heard you had a good sense of humor!" We all started laughing. This was originally a Negro joke that had been picked up by Native Americans. Interracial joke-sharing: the world is changing.

A Reservation Encounter

This story begins on a very bitter cold Montana Saturday morning. It was so cold the snow screamed underfoot as I hiked on the Rocky Boy Reservation south of Havre. I loved the reservation for its one tall mountain, Baldy, which rose above the thickly forested pine slopes crisscrossed with trails. The summit was often reserved by the local Chippewa Cree, who climbed there for vision quests or other ceremonial traditions. I respected this sacred summit and only climbed there when the local tribe declared open days for anyone to hike. But the lower trails were always open to everyone.

On this particular day, I drove out some twenty miles into the Bears Paw Mountains. I parked on the side of a half-plowed road near a trailhead. It was a relief to be out in raw nature after a week of inspiring high school French students to use the subjunctive verb forms. I was bundled up in layers of coats, scarves, snow pants, hats, gloves, and boots. I hiked steadily, breaking trail through a few inches of snow. There wasn't much because it was too cold to have snowed more. After a couple of hours, I stopped and was hurriedly eating a sandwich before heading down. In the silence, I heard the crunching of someone coming up the trail behind me.

I turned and looked down the slope. A native man from the reservation was following my footprints, his head down, his hands jammed in his pockets. I cleared my throat so he wouldn't be started when he saw me. He looked up and said, "Oh, hello. What are you doing out here on such a cold day?"

"Yeah," I replied. "It really is bitter, but I had cabin fever and just wanted to get out and enjoy nature. What about you?"

"Well," he said, "I am on a mission. My mother died a few weeks ago. She is buried in the Catholic cemetery, but I wanted to honor her memory in a traditional way. I asked an elder how I might do this. He said that I should get some long lengths of blue and white material, her sacred colors, and tie them to a special tree up on the summit. I did that last week. Today he told me I should go up to the summit, untie them, bring them down, and place them on her grave. So that is what I am doing."

"I am sorry to hear about your mother," I offered. "Is there anything I can do to help?"

He hesitated, then said, "Yes. One thing. My friends drove me out here. You will probably see them down on the road. Please tell them I am fine, but to be sure to wait for me until I get back down."

He turned and started back up the trail as I headed down. I did find his friends and relayed his message. I felt moved by this unexpected encounter, and privileged that he had shared with me this significant act of traditional respect.

Hutterites

Native Americans are not the only minority peoples in northern Montana. There are several colonies of Hutterites near Havre. They are members of an anabaptist movement who left Europe to escape prejudice there. They live in colonies of one hundred or so members, dress in traditional garb, and speak an old Germanic dialect. Unlike the Pennsylvania Dutch groups, the Hutterites use the most modern farm machinery and agricultural methods. The local Montana farmers criticize them because they have bought up smaller farms, are exempt from individual income tax because they own everything communally and are pacifists. However, they always give blood at the local blood drives and pay corporation taxes. My husband and I enjoyed visiting their colonies, and we were always welcomed because we spoke German with them and took an interest in their lives.

Postscript

By 1989, my husband and I, as well as our sons, had all left Montana for university and work. My husband had a contract to teach international relations in a master's degree program for Air Force officers in Europe through Troy University. I taught English as a foreign language in Germany and then in the former Eastern Bloc countries after the fall of the Berlin Wall. We saw the prejudice of the West Germans toward the East Germans for not being as up to date as the Westerners were. Later we spent time teaching in Korea, where the Japanese were still hated. Prejudice, for whatever reasons, seems universal.

As I write this now, I am back in Boulder and a member of the NAACP, and we are still dealing with prejudice and inequality. I am grateful I have had so many opportunities to live with diversity in many forms as I continue to do what I can to make Martin Luther King Jr.'s "I Have a Dream" speech come true.

No Home Base—Nothing Permanent but Taxes

From 1988 until 1998, Hajo and I had no home base. We were academic nomads, teaching at military bases across Europe in the case of Hajo, and in my case, teaching business English in Eastern Europe when the Soviet Union collapsed. In fact, because we had no home state of residence, we could not vote in US elections, though we still had to pay our taxes!

Though we spent much of this decade in Europe, it was not the first time we'd lived abroad together for an extended period. Hajo, the boys, and I spent a sabbatical year in Berlin from 1979 to 1980. Eric and Mark attended the International John F. Kennedy School, which, incidentally, Pop had helped with accreditation when it was founded. They thrived in the city atmosphere. At first, they didn't want to leave Havre for a whole year, but when it came time to return home, they wanted to stay in Berlin. We had the opportunity to attend their first full-scale rock concert, a Supertramp show, as well as experience other music offerings and museums. By 1988, both boys had graduated high school, and we were ready for the next phase of our adventures.

Living in Divided Germany

When World War II came to an end, the major powers of the European conflict met in Yalta to divide defeated Germany. Germany was separated into four distinct zones: American, British, French, and Soviet. Berlin, the traditional capital located in the center of the Soviet zone, was also divided into four sectors. In 1961, a wall was hastily constructed by the Soviets, encircling West Berlin (American, British, and French sectors) from East Berlin and East Germany. The Soviets declared the wall necessary to control Western spy activity.

The famous Checkpoint Charlie was the official crossing point from the American to the Soviet sector. Allied Forces members could cross into East Berlin after a thorough examination of passports and sometimes other documents. However, to reach Berlin in the first place, the Allies had either to fly or cross through the Soviet zone of the country, which was delineated by guard towers, barbed wire, and a no-man's-land. The term Soviet was used by the Russians to indicate that East Berlin and East Germany had a degree of independence as part of the Eastern Bloc of nations.

Most people flew into Berlin to avoid the confrontations of driving through the Soviet zone. One day, Hajo, our two sons, and I drove through the Soviet part of Germany to reach Berlin. We stopped at a filling station restaurant for lunch. We found a parking place that seemed legitimate. We had been told that the East Germans liked to ticket cars for parking illegally. When we returned to the car after lunch, there was a ticket on our windshield for a huge sum. The Vopos (border police) were waiting for us. We said we had looked and saw no indication that this was an illegal spot.

Living in Divided Germany | 137

They showed us a 3x5 inch card on a pole facing away from the street that clearly stated no parking. We should have seen it. We had no recourse. We had to pay the stiff fine in West German currency, which was worth four times more than East German marks at that time.

This was a small indication of life under communism. Before the war, Hajo had had family and friends in what would become East Berlin. When the wall was built so quickly, all relationships were cut off between East and West. Only nationals from the US, Britain, or France could travel freely into East Berlin. I became an informal courier for clothes and family items that had been left when East Berliners made a hasty move to West Berlin. I remember wearing winter clothes from East to West for a friend who had escaped to the West, leaving behind home, job, and belongings. Fortunately, the West Berlin government helped her and many other "refugees" find what they needed.

We had close friends in East Berlin. Werner was an artist who had not been able to leave the East because of his elderly mother. The Communist regime demanded that he join the Artist's Union based on socialist realism. The officials told him that if he didn't join, he would have two options: to repay all his former education expenses or face being jailed. As a matter of principle, he refused to join the union, but he did not have the financial means to repay what the government demanded.

Some of us in West Berlin contemplated lending him money. One of our friends said he would help Werner and sneak money to him, but at the last minute our friend changed his mind. When our friend crossed through the control point into East Berlin, he was given a complete body check and searched for bringing in illegal goods. Somehow the Vopos suspected he would be bringing something. Fortunately, nothing was found because he had changed his mind. After this close call, Werner said he had found a way to repay the officials and that we should not endanger ourselves.

Our friendship with Werner continued. He gave us a pre-wedding party in East Berlin since he couldn't travel to the West for our wedding. (Incidentally, another East German friend had tried to get permission to attend her father's funeral in West Germany. They granted permission, but only a week after the funeral was over.) Werner gave us a fun party, inviting my parents and brother and my sister, Jane, who made friends with his daughter at this party. She was very upset to learn that her new friend could not stay in touch with her or travel to West Berlin.

A few years later, Werner showed particular interest in our camping trip to Russia. He asked us all kinds of questions about it and why we were going. Hajo explained that it was part of his position teaching international relations, which included Russia and the Soviet Union. Several years after that, Werner was sent to the United Nations in New York City as part of an official delegation showcasing books from East Germany. Hajo and our son Mark met him and showed him around New York City. Werner was very nervous and kept looking over his shoulder to see if they were being followed.

Over the years, Werner developed a drinking problem and became obese. We thought it was due to the pressures of living under a totalitarian regime. Finally, a few months before the wall came down, we received a notice that he had died of a heart attack. We sent a letter of condolence but heard nothing from his family.

We continued living in West Germany. Unexpectedly, we heard people were dancing on the Berlin Wall. A new era opened before us. Sometime later, we received an official notice from the security office in Berlin that our record from the Stasi (the East German secret police) had been found, and we could have a copy if we wished. We knew nothing about having a spy record in East Germany. The copy arrived with great black splotches covering up information about people other than us. We discovered that Werner had become an informant and created a false record of our actions as spies from the West. He wrote outrageous stories, but nothing that would really cause us problems. According to our record, I had to be an undercover agent because I asked so many questions. My father was said to be a special friend of famous US General Lucius Clay, and to speak many foreign languages. Hajo was said to have constant and close contact with the US military, but there were no details.

Hajo was distraught when he saw the record because he felt Werner had betrayed their friendship. I was deeply saddened that Werner had been so manipulated by the Communist regime, but finally we understood how he had repaid his debt. He had no other choice. We lost touch with the family. We knew his daughter had defected to West Germany years earlier. His wife never responded to our letters.

1989 and a Reunited Germany

Euphoria reigned during the first months of the reunification of Germany. In December, we joined people climbing on top of the parts of the wall that had not been torn down, celebrating wildly and collecting pieces of it as souvenirs. Beethoven's Ninth Symphony was conducted by Leonard Bernstein with the chorale word *Freude* (joy), changed to *Freiheit* (freedom).

Our son Mark and I joined the crowd of runners on the first West–East Berlin Run through the Brandenburg Gate and the wall on New Year's Day 1990. The Vopos did high fives with us as we crossed through the broken wall. Weeks earlier, they would have shot people running into West Berlin. I was invited into East German high schools to talk with students about life in the West. It was a time of meaningful sharing about our mutual hope for the future. East and West were united in celebration. Gradually, though, the tough realities of unification dulled the euphoria.

Shortly after unification, I began teaching business English as a foreign language to unemployed East German professionals. City University, based near Seattle, had a contract to teach business English to prepare unemployed professionals in former East Germany to study for an MBA. Many of my students were bitter about being unemployed. Under Communism they'd had guaranteed jobs for life. But they were happy they did not have to wait years to buy a Trabi, the East German car, and that they could travel to West Germany, but they had very little money. One student declared to me that if he couldn't get an MBA and a good job, he would go to Iran and use his East German technical chemical degree to help Iran defeat the West. His attitude was the exception, and most students were optimistic, looking forward to the opportunities ahead.

I remember, however, that those older than fifty didn't feel they could start over. One elderly East German man bemoaned the new life. There had been no traffic jams under Communism—the police knew how to direct the traffic (there wasn't any). He couldn't find the kind of delicious local sausage he used to buy, which was no longer available. The white roses he grew in his garden were no longer available, either. Prices were constantly rising, but his pension was the same as before and much lower than pensions in West Germany. Meanwhile, in West Germany, where I also taught English classes at Siemens, many of my students said that the workers coming from the East lacked the skills to work in the West; they were years behind. Why should West Germany be helping them out?

Some young people from East Germany had serious problems. Hajo and I lived in an apartment in West Germany. Two young former East German workers rented another apartment in the same house. We could see they had a drinking problem by the quantity of beer and liquor bottles in their trash. Our landlady said they were having a hard time paying their rent. She was very lenient. Meanwhile, Hajo and I were both transferred out of the area for work assignments. When we returned several months later, we asked the landlady about the two young men who were no longer around. She said, "One disappeared. I found the other had hanged himself in the shower. I called the police. It was terrible. They just couldn't adjust to the changes of a united Germany."

Many East German teachers were excited about their opportunities, though their pay was considerably lower than that of teachers in the West. Hajo and I worked with a German educational institution to organize a monthlong English teaching seminar for them in Troy, Alabama, with Troy University, Hajo's employer. The participants were all English teachers from Saxony in eastern Germany.

We flew together into Alabama and found our lodging on the Troy campus. It was a Sunday evening. The teachers thought they would like to relax with a beer, but no beer was available in Troy on Sunday. They asked around and learned that beer was for sale in a store just outside the town limits. A couple of girls and guys volunteered to walk for beer! They started along a road heading out of town.

This was the South. People rarely walk in Troy. A police car happened along. The officer saw this group walking and pulled over to investigate. The Germans felt a bit intimidated by a policeman stopping for them, but he seemed friendly. He heard their foreign accent and asked if they were lost. They replied, "No, we are just looking for beer for our group. We are from Germany."

The officer said, "It's a long walk to the store. Let me give you a ride. We can't have thirsty Germans around here!" The policeman drove them to the store, where they bought the beer, and then he deposited them back at the campus dorm. The German teachers never forgot the kind American policeman!

I cannot write about all my experiences in Germany, divided and united. But I love the natural beauty of the country. I love the forests, and the Volks Marches, which are organized walks of ten, twenty, or more kilometers through forests with marked paths so hikers don't get lost. The marches finish at refreshment tents with beer, coffee, and cakes. I love the vineyards and the variety of wines and the opportunities to join in the grape harvests. I have picked grapes often in the Mosel River Valley with its steep slopes and curved river passages, and in the Rhine Valley in the vineyard of the abbey where Hildegard of Bingen lived in the Middle Ages. I pretended I could hear the sisters chanting the hours as I picked the grapes and the sunlight moved across the rows of vines as it had for centuries. I love the Bavarian Alps, where I hiked across flower-dotted meadows in summer and cross-country skied in winter. I love the village church bells ringing the Angelus and calling people passing by to stop and listen at each sunset. I love Germany for its great classical music and operatic contributions, which so enrich the world!

I love Germany because it is in my blood. The family on my father's side immigrated to America from Germany in the nineteenth century. My husband was born in Berlin and can trace his German roots back to the sixteenth century. Our granddaughter majored in German in college and now works with German law enforcement through Amazon. Many of my best friends are German, and we keep up our friendship across the continents. I celebrate Germany's many positive aspects and hope that as a human race, we can learn from and move beyond the violent and tragic realities of its past.

Figure 22. Mark and I waiting to run the 1st West-East Berlin Run, January 1, 1990

Mother Teresa

Even though Germany is quite a secular country, Germans celebrate a Roman Catholic Week and a Protestant Week on alternating years. Our family lived in Berlin during the year of a Catholic Week, when church leaders and listeners gathered to discuss world issues in the context of faith. Strains of "dona nobis pacem" echoed through the subway tunnels while Catholic youth groups shared bread and wine in a spirit of reconciliation in parks and on city sidewalks.

Mother Teresa was the celebrity of Catholic Week. Much of the world considered her one of the world's outstanding women, and she received the Nobel Peace Prize. She was admired for serving the poorest of the poor alongside the sisters of the Missionaries of Charity, which she founded.

She was scheduled to speak to a gathering sponsored by Catholic women in Berlin at a large convention center. I arrived at the event just as she was being escorted through the lobby to the auditorium. The lobby was crowded with hundreds of robust German women bedecked in furs on a frigid day. The contrast between these Valkyries and the diminutive woman in a light sari and sandals was startling!

She seemed even smaller behind the speaker's podium. I thought she might be intimidated by the mass of Aryan energy before her, but she spoke simply and forcefully of her work in the streets of Calcutta with the poor and dying. She told the story of an elderly woman brought to her by the missionary sisters. The woman was complaining bitterly that she had suffered so much in her life, that she found it unfair and couldn't understand why she had suffered so much pain. Mother Teresa suggested to her that God must have loved her in a special way to have chosen her to suffer as Jesus had suffered. God was indeed with her to help her in her affliction. Furthermore, she need not fear any more suffering because she would be cared for by the sisters for the rest of her life.

Her theology was often criticized, but Mother Teresa's care of the poor and dying spoke more eloquently than her words. Despite the critics, her example was one of the inspirations for my own work later with the poor of Haiti.

Two Months in Bratislava

When the Berlin Wall fell in 1989, the repercussions were felt all over Europe. The Slovnaft oil company in Bratislava, Czechoslovakia, was a terminal point of a Russian oil pipeline. The managers of Slovnaft wanted to improve their English skills in order to better communicate with their Western counterparts. Instead of preparing students to follow MBA courses in English, my stint in Bratislava was to help company managers actually communicate in English in the business world. This was a daunting task in two short months before Easter in 1992. Meanwhile, the country itself was in the process of dividing into two entities: the Czech Republic and Slovakia. I was in the midst of the change from a Communist state to two democratic countries.

My living situation in Bratislava was old regime. The Slovnaft company set me up in a company apartment in a cement building on the fifth floor. I reached it by dark stairwells or by an infrequently working elevator. Broken venetian blinds hung haphazardly over the windows, as well as over a glass door leading to a small balcony from which to see what was happening in the neighborhood. I was paid the standard equivalent of fifty dollars per month, along with free rent, a free noon meal in the company canteen, and the most kind, helpful students imaginable. The food was usually dumplings with gravy and a small piece of meat, and maybe cabbage. Twice a week, I walked to the local train station to buy fresh radishes from a farmer, the only fresh produce available in March when I was there.

The students showed me how to obtain symphony and opera tickets for fifty cents each, until concertgoers from nearby Vienna heard about the inexpensive tickets and bought them up. I remember in one of the concert venues, the Czech concertgoers sat on one side of the auditorium, the Slovak on the other. I asked why and they said, "Custom!" The country was already divided before the official division.

One evening, I was in my apartment when I heard a commotion outside. I went out to my balcony and saw a fire truck spraying water up to the seventh story and into an apartment a few balconies from my own. Thick smoke was pouring from the apartment. My first thought was to evacuate the building—grab the important papers and valuables and leave. But no one was evacuating the building. We were all on our balconies watching the activity around us. Firemen entered our building and climbed up the stairs to the burning apartment. Soon there was no more smoke, and everyone watching from the balconies went back inside. Since the whole building was cement, the fire did not spread, and most of the firemen left. I couldn't find anyone who spoke any language I could understand to learn what had happened. It was all in an evening in Bratislava!

Volunteerism also played a role in Bratislava. One of my adult students was a leader for the regional Paralympics. He organized fundraisers and competitions for young athletes with disabilities. He invited me to participate in a five-kilometer group fundraiser run. It was special to participate in the run with people of all ages and abilities in this newly emerging democracy.

The Gospel according to Mary

One of my students asked me if I would like to attend the rare performance of a folk opera. For a number of years in the former Eastern Bloc, the Communist government had allowed local amateur theater groups to present *The Gospel according to Mary* during Lent. It was a musical originally composed and produced in Hungary. Its popularity took it to other countries, where it was translated into the local languages. Only a few performances were permitted, and all participants had to be amateurs. Tickets were almost impossible to obtain, but my student had an extra one and thought I might be interested. I was!

I was incredulous that a totalitarian country governed by the ideals of socialist realism had allowed such an obviously Christian public performance. I was attending the first production since the Fall of Communism. I had the feeling that the audience was remembering with nostalgia a special event of the "good old days." The audience in the traditional theater was composed of old folks, families, and young professionals. The story followed the Biblical action in the life of Jesus, but from the point of view of Mary, his mother. Each scene used a different type of music, from classical forms to jazz, from soloists to large group ensembles. The staging was creative, from shepherds on bicycles to aerial inventions. It may have been an amateur production, but the superb quality of the voices and the dramatic acting was totally engaging.

It begins with the annunciation. Mary, a young woman, is alone on the stage when suddenly a deep voice booms out "Hail, Mary . . ." It is the angel Gabriel. He is not alone but is accompanied by a full angelic host singing the Magnificat in the Russian male-voice tradition of Orthodox church music. It was fantastic. Little wonder the message included a gripping "Fear not!" The juxtaposition of timid Mary with this display of heavenly power was dramatic.

The folk opera continues with many other scenes, some funny, others touching. At one point, Mary is teaching Jesus basic words. In a simple duet, Jesus learns and repeats the words that are emblematic in his later life: water, bread, wine, seeds, pearl, stones, and others.

Throughout the opera, the figure of Death plays a major role, portrayed as a tall woman clad in black. She sweeps onto the stage accompanied by dissonant chords from the orchestra. She seeks to kill the infant Jesus in the tradition of the Massacre of the Innocents. The holy family escapes from her into Egypt. Death tries to lure Jesus away from the temple as a boy while Mary searches for him. Mary sees people trying to stone Jesus for his teachings. Death motivates them, with dark, whirling intention. Each scene has a different type of music, but it always includes the minor chords of death.

Finally, Death believes she has succeeded in killing Jesus in the crucifixion. She gloats over her accomplishment, draping herself over the empty cross which has been lowered to remove Jesus's body. In her demented joy, she is alone, unaware that the cross on which she is leaning is slowly rising from the stage, carrying her with it, ever higher. In her last soaring note of elation, she suddenly loses her balance, plummets from the cross, and crumples lifeless to the ground. Death itself is dead!

Musical theater in its various forms can portray the familiar story with dramatic proportions and an immediacy and intensity that words alone cannot convey. I am thankful to my student for the ticket! There were numerous other oratorios I also enjoyed, all leading up to Easter. Meanwhile, local artists were decorating Easter eggs of many colors with intricate white designs and wooden carved rabbits pulling wagons for the eggs. Lent was a creative time to be in Bratislava!

In Conclusion

The Slovnaft directors decided that on my last day of teaching, there would be a party in English. It would include the five vice presidents, all male, with whom I had been working the most. Instead of the schedule, they would take off the entire afternoon. The thank-you party was elaborate, with beautiful gifts of hand-painted pottery and free-flowing Becherovka, the local liqueur of the area. We started in English but were soon toasting in every language we found in common: Czech, German, French, Russian, and more. They were genuinely appreciative, not only for their own classes, but for the ones I had taught to other workers in the company.

I could not imagine the top executives of a Western country taking off an entire afternoon for such a party. I had met an American executive of Texaco who was also working at Slovnaft to help make their company competitive in the West. He said they would have to fire at least half of the staff as a starter. The Slovnaft management was unwilling to even consider it, since one of the Communist practices was to hire workers for life. Eastern Bloc people willingly accepted the Western values of free speech, and many other things, but the economic challenges of competition and capitalism were overwhelming for some.

Poland Is Open

One result of German reunification was the opening of Poland to the West. We met Irena and Tom, both Polish secondary school teachers, at an international educational conference, and they became close friends over the years. They invited us often to their summer house near Gniezno, and we invited them to go camping with us in the American West. They loved the West, from Yellowstone geysers to cowboy hats, from picnics in the Bears Paw Mountains near Havre to wading in the Great Salt Lake. They reciprocated, taking us to see the reconstructed city of Gdańsk (Danzig) in all its eighteenth-century glory! We walked together many times through the streets of Gniezno. On one such walk, they took us into the historic cathedral, the site of the crowning of numerous Polish kings. What I remember most about this cathedral had nothing to do with kings!

Father Maximillian Kolbe

In the cathedral, Irena and Tom led us up a shadowed aisle to a simple side altar. The scent of incense from an earlier mass still lingered in the air. We stopped by a pillar where several candles were flickering, lighting a small image attached to the stone wall. It showed an emaciated man, head shaved, wearing the striped suit of a Nazi concentration camp prisoner. A small sign under the picture read, "Maximillian Kolbe." We all stood there in silence. We each knew his story, but I didn't expect to see this small altar to his memory in the cathedral of kings.

Father Kolbe was a Franciscan priest who was sent to Auschwitz. He had been arrested for harboring scores of Jews in his monastery. Survivors of Auschwitz testified that he brought great comfort to his fellow prisoners of several faiths and often gave his meager rations to others.

One day at roll call, one prisoner was missing. According to the rules at Auschwitz, ten prisoners would be executed in place of the one who had escaped. Ten men from his cellblock were immediately named to be put to death. One of those victims, a young man named Gajowniczek, cried out, "Oh my poor wife, my poor sons. What will become of them?"

At that moment, Father Kolbe, his cap in his hands, his head bowed, stepped forward and said, "Please take me in place of him. I am old and he has his life ahead of him."

The guards hesitated a moment, then agreed to execute Father Kolbe in place of the young man. Thus, Gajowniczek was saved.

For the next two weeks, the ten victims were starved in a dungeon. None of them knew that the supposed escapee had later been found dead in a latrine. Finally, only Father Kolbe and three other men were left. These four were given lethal injections that ended their ordeal. A witness said that Father Kolbe put out his arm and gave a whispered blessing to his executioner. He died at age forty-seven.

Gajowniczek lived and was freed from Auschwitz by the Allies at the war's end. He returned home to his wife, but his two sons had died. Every year he made a pilgrimage to Auschwitz in memory of Father Kolbe until his own death at ninety-two.

Today, Maximillian Kolbe is venerated as a martyr and saint. We four left the cathedral and continued our walk, moved by the sacrifice of Kolbe, happy that we could share friendship in a new world of reunification.

Grand Glimpses of Portugal

In elementary school, I learned about the great early explorers, which whetted my appetite for exploring the world myself. Many of the first were Portuguese, such as Vasco da Gama, who sailed from Portugal around Africa to India! Centuries later, Hajo and I travelled from Madrid to Lisbon in February 1989. We left the high, frigid plains of Spain for the early spring of Portugal, filled with the fragrant blooms of open flower-markets. We had the good fortune of finding a tourist-taxi driver who took us immediately to the Discoveries Monument. There we saw and celebrated how world exploration began with Prince Henry the Navigator, da Gama, and Magellan. I learned the latter had died of a poisoned arrow during a battle with local native tribes in the Philippines on April 27, my birth date. This fact has not discouraged my own motivation to explore the world!

There is so much to love about Portugal. We discovered the Algarve with our Dresden friends, who were always eager to see the world denied them during Communist times. Its coastlines are spectacular, with golden cliffs and caves and turquoise waters. We loved exploring on the picturesque trails and stairways, but the seas were too rough for kayaks or boats when we were there. I was impressed by the lovely tiles on house facades, and those for sale in many forms. I also enjoyed the light, refreshing Vinho Verde wine.

I especially appreciated the two months I spent with Hajo when he was teaching in the Azores islands for the US Air Force. The Azores have been considered part of Portugal since the fifteenth century, even though they are far off the Portuguese coast in the Atlantic Ocean. In summer, we drove the winding roads, lined with blue hydrangeas as far as the eye could see. The fields were a patchwork of green and gold, like giant quilts spread over the undulating hills. Here and there we hiked to small lakes that filled the craters of ancient volcanoes, reminding us of the volcanic origin of these islands. The highest point in Portugal is Mount Pico, a volcano on Pico Island. I wished I had climbed it, but the weather did not cooperate. We went whale watching instead and followed an awe-inspiring sperm whale, the largest living animal I have ever seen!

The villages on the various islands are made of whitewashed homes with red tile roofs, their windows outlined with dark paint. Each village has its own small marching band to play for the numerous festivals. As we watched the processions, we were reminded of the marches of John Philip Sousa, who was born of Portuguese parents who encouraged his musical career. Each of the Azores has its own feel and special attractions, including windmills with cloth sails on Corvo and a plethora of flowers on Flores. They are all connected by Portuguese ferries, which I will admit can provide some very rough crossings. But all the islands convey a unique touch of Portugal far from their home base.

Zanzibar

Hajo and I had never considered Zanzibar as a probable tourist destination, but twice, in 2003 and 2007, it was part of tours we made to that part of the world. On one tour, we were given a free day to explore as we wished. Thus, we found ourselves on the east coast of Africa in this ancient sultanate and trading center, full of history and intrigue. These trips occurred after Zanzibar became part of Tanzania, but it still retained elements of its varied past in its architecture and its racial and cultural diversity.

The first site we visited was an Anglican church built over the ruins of an eighteenth- and nineteenth-century slave market. In a windowless dungeon, slaves were put in irons awaiting sale and eventual transport on ships to the Americas. These Black Africans were brought to Zanzibar from the interior of Africa. Many were taken to the South in the United States, but the majority went to French and Portuguese plantations in the Caribbean. These slaves had often been captured by Arab traders, who had been trading and using slaves themselves for ages across the Islamic world. Even through the first part of the twentieth century, slavery continued in parts of the Middle East, and it was only declared illegal through international pressure. It was our trips to Zanzibar that made me aware of how common the practice of slavery has actually been.

One of our guides asked if we knew the hymn "Amazing Grace." We did. Our guide asked, "Did you know it was written by a slave trader shipping slaves from Zanzibar to the Americas?" We didn't. Since then, I have traced the history of this hymn by John Newton. He was returning to the British Isles from a slave-trading trip to the Americas when a terrific storm threatened to capsize his vessel. In a prayer of desperation, he called out to be saved. This was his first time of conversion. However, he continued in the slave-trading business from West Africa, not Zanzibar in East Africa, for five more years before he fully realized the evil of his ways and became an abolitionist.

During our second trip to Zanzibar, Hajo and I planned to walk peacefully through the old city, exploring on our own. Fortunately, we were followed by a tall Black citizen of Zanzibar who told us he was the best tour leader in the area. He proceeded to give us facts about the buildings we were admiring, and we hired him for the day.

He explained he was part of the Muslim minority in Zanzibar, some 35 percent of the population, the majority of which were Christian. It was the day after the end of Ramadan, and everyone had been celebrating overnight. That was why there were relatively few people on the streets. He was expecting his two daughters to meet him further along. Two cute young Black girls in pink dresses came running when they saw their father. They talked excitedly and then greeted us politely in English before they skipped away back down the street. Their dad said their mother wanted the whole family to gather at home in a couple of hours, so he would have to cut short his tour with us.

During our tour, he told us part of his life story. He grew up in Zanzibar in a traditional Muslim family. He had been taught that Islam is a universal religion with equality for all. As a young man, having finished a university degree and being fluent in Arabic, English, and German, he went to Egypt to find a job. When he arrived, he registered with various employment agencies and companies, expecting to find work with his excellent credentials. Instead, he was mostly ignored and turned away. He discovered that even though he was seeking employment as a Muslim in a Muslim country, he was too black. After a year of numerous rejections, he returned to

Zanzibar, where he became an official tour guide. Humanity around the world has a long way to go to achieve racial equality.

Sister Mary: Witness and a Catalyst for Change

While I was enjoying my sabbatical year in Berlin from 1979 to 1980, the rebels in Rhodesia were attempting to bring an end to colonial rule in this African country. They attacked various British establishments, including a mission school under the direction of a Catholic religious order in Ireland. Meanwhile, in Ireland, the Protestants and Catholics were also at war.

I did not really relate to these issues personally until I decided to spend a week during the spring of 1980 in Taizé in eastern France. Taizé is an ecumenical retreat center that was established in 1940 by a Protestant brother named Roger Schutz. During the war, it served as a center of refuge and as a route for Jews to enter neutral Switzerland. Since then, up to one hundred Catholic, Protestant, and Orthodox brothers have joined the ecumenical order. They have developed a style of worship and music that has been adopted by many churches in the United States. Taizé draws thousands of students from around the world to meet, worship, and discuss world issues in a spirit of reconciliation.

I was among a group of older participants. I was frankly a bit intimidated when a rather robust, austere-looking nun in full black garb sat by me. She whispered in heavily Irish-accented English, "Sorry to bother you. Can you help me? I don't speak French. I just arrived here. What is happening?" I explained as best I could as some chanting began. After the service—which included Communion at two different altars, one Catholic and one Protestant—I helped her register and find her room. Then she told me her story.

Sister Mary: Witness and a Catalyst for Change | 161

She was on her way from Rhodesia back to her order's mother house in Ireland. Her superiors had routed her through Taizé, thinking she would appreciate the spirit of the community. She had been teaching for many years in a mission school in rural Rhodesia. Several weeks before, she had been relaxing at the beginning of a school vacation. The students had all left, and she was there as the head of a group of her teaching nuns. A number of priests were also in the mission compound.

She recounted being in the shower washing her hair when she heard machine gun fire just outside the compound. She immediately ran from the shower, wrapped herself in a towel, and rolled under her bed as the gunfire continued. Lying there on the floor, her hair still lathered with shampoo, she wondered what she should do. She said she prayed weird prayers—that God would save her because her family back in Ireland would be prejudiced against the African Blacks if she were killed!

As suddenly as the gunfire had started, it stopped. She quickly dressed and checked on the other nuns. They had also gone under their beds. Physically they were fine, but they were emotionally shaken! Then came a knock at the door. It was the head priest coming to check on them. One of the brothers had been grazed by a bullet, but it was nothing serious. He was lucky! The priest said they would have to leave the mission. They were too vulnerable out in the countryside. But it was too late to try to drive away that day. They would have to wait until morning. He was sure there would be more rebel attacks in the night. They would have to stay down, turn off all lights, and keep away from windows.

The nuns decided to barricade themselves in one room, barring the window as best they could and lying on the floor on mattresses. Sister Mary continued her story; I will tell it as best as I can recall it. "Sure enough," she said, "as soon as it got dark—and remember, this was darkest Africa," she added with a twinkle in her eye, "firing began again from outside the compound. We sisters huddled together on the mattresses and heard machine gun fire on all sides of the compound, sometimes nearby, other times further away. It continued off and on all night. When dawn began to glow on the horizon, the firing stopped and all was silent around us."

The nuns and priests were all safe. This was due to the incredible action of a priest of German background. As a young soldier he had fought in the Wehrmacht, and he was somehow able to leave Germany and move to Ireland with a concealed weapon. He had become a priest in his new country. When he was assigned to go to Rhodesia, his superiors, who knew of this weapon, obtained ammunition and sent him off with his armaments. All night long he had run from one side of the compound to the other, returning fire and firing off rounds himself to give the impression that it was well guarded. Sister Mary said, "He prayed continually that he would not accidentally kill a rebel. He just wanted to keep them away. His ploy worked. We all drove away from the compound, I with shampoo still in my hair!"

Sister Mary and I sat together during many services. Each day, she reached out to all the English-speaking students and participants, inviting them to meet her for afternoon tea in a lounge. She said it was the only thing she knew how to make, and she always had her teapot. Every day she had a following of a dozen or so tea drinkers, mostly students from Ireland, both Catholic and Protestant, who had neutral ground to discuss the violence in their country. I witnessed some deeply moving encounters as they shared their experiences and hopes for the future. They said they hoped that someday they could take Communion at a common table. But for now, Sister Mary said, "We can drink tea from a common kettle!"

At the close of our sessions at Taizé, I accompanied Sister Mary to her train in Paris, which took her on to England and then to Ireland.

She later sent me a book written by a priest who had walked from Ireland to Rome. "Not across the channel," she added. "He wasn't that saintly!" After that, I heard from her no more. But I have a new appreciation for the faith, good humor, and spirit of reconciliation she exemplified! Time has passed and peace has returned to Ireland, and Rhodesia has become Zimbabwe. This is but one story of what happened in the process.

The World Is 70 Percent Water

An Astronomical Event—January 2007

Hajo and I were off the coast of East Africa on a small cruise ship in the southern hemisphere. I had just finished dinner and slipped outside to walk along the deck for my evening constitutional. Wow! I stopped dead in my tracks. There in front of me, a huge comet was streaming across the sky like a white-feathered pen dipping into the inky sea. This great arc of light commanded my attention! It was a magical apparition. Where had it come from? Why hadn't we heard of it? It dominated the dark southern sky!

Other folks from the cruise walked out on the deck, gave it a quick glance, and reentered the ship for the evening program. A few others stayed with me on deck. We were all mesmerized; this was the best evening program you could wish for. The comet gradually slid further toward the horizon. The white orb was swallowed by the ocean swells and then was no more. Later we learned that this was the McNaught Comet, first discovered some time before by an Australian astronomer, the second-brightest comet in decades. The next night, I looked for the comet again. It was a dim blur on the horizon. The third night, it was gone.

Cruise with Kir

Hajo and I enjoyed cruising for many reasons. We loved to be waited on and enjoy gourmet meals. On one cruise, we received free wines with dinner. The white wine was a French Burgundy called Aligoté. I was transported back to my days living in France, where this wine had a special connotation. It was served with a splash of sweet black currant liqueur (cassis) in a drink called kir. Kir became a popular drink in many countries, but no one seemed to know its origin. One of my French friends told me the particulars of the French priest for whom it was named.

My friend lived in Dijon, where Félix Kir was first a priest and later a canon in the cathedral there. He had been active in the French resistance, served as mayor of Dijon for many years, was personally invited to Russia by Khrushchev, and received many honors and awards. As a parish priest, he had little personal spending money. He would stand by a bus stop until he saw a bus containing one of his parishioners, and would gesture to him and invite him to a local café for a drink. He would always order a glass of white wine, usually the local Aligoté, with cassis. He wouldn't have money to pay for it, so his guest would oblige. It got to be a local joke.

A story is also told about the time Kir got into a discussion with an atheist who asked him how he could possibly believe in God. He retorted, "Monsieur, you have never seen my rear end, but I assure you that it exists!" This was the remarkable, earthy man for whom kir was named. Hajo and I toasted his memory as we enjoyed the Aligoté with a splash of cassis!

A Rowboat Rescue

In December 2011, Hajo and I decided to make a repositioning cruise, which meant making a voyage from Europe to the Caribbean when the Mediterranean cruise season was over. We boarded our ship in Portugal and set sail across the Atlantic. It was basic cruising: no programs, no lectures, and no off-boat excursions, though there was an excellent Russian string quartet that played for several hours a day, and the ship was colorfully decorated with Christmas displays. One evening after dinner, I went out on deck to see the evening moon rising over the port side of our vessel. It was cold, so I hurried down to our cabin to put on a coat and hat. When I returned to the deck, I was surprised to see the moon above the starboard railing. What was happening? The moon does not move that quickly. The boat had to have turned 180 degrees, but why?!

Later, I found a ship's officer. He explained they had received an SOS call, and being the nearest ship, they had turned back to rescue a crew of two young men rowing across the Atlantic. Their rudder had broken, and they were at the mercy of the sea. They were part of a rowing competition and had been making very good time until their mishap. The next morning at breakfast, I talked with them. They were grateful to have been rescued, but disappointed at having to give up the race when they were doing so well. I had no idea there were rowing competitions across the Atlantic. They used special sculls where they could both row together, or one could row while the other stretched out and slept. Their rescue was the highlight of our cruise!

Church in the South Seas

Over the years, Hajo and I made several cruises in the South Seas. We had all the typical cruise amenities and others on the side. The unexpected moments are what I like to remember. In Tahiti, we attended a local Sunday church service. We expected that the service would be in French, the official language of Tahiti. We arrived at a white-frame church with slanted wooden shutters that could be opened to the breezes or closed when it rained. Church ushers greeted us warmly at the door and then asked what we sang: soprano, alto, tenor, or bass-baritone? I sang soprano, while Hajo sang bass. We were seated accordingly—not together, but according to our voice.

The church was divided into four sections, and all the numerous hymns were sung in four parts. Most people knew the hymns by heart and sang their respective parts in the local language, not French. We hummed happily along. All the women attended in white dresses, with white hats to protect them from the sun. The men wore light-colored suits. It was a joyous setting and service.

Biking in Bali

Hajo and I made two stops in Bali, each for one day, to do as we pleased. The first time, I rented a bike and rode clockwise around the island, the second time, counterclockwise. Both times were fun. I stopped to examine ancient carved stonework in various places. One piece was a large stone turtle. Many islanders regard a turtle as the traditional origin of life. I rode close by a church where I was chased by a barking dog. The priest appeared at the door and called to me, "Don't worry. The dog doesn't eat Christians!" But the priest leashed him, and I was safe, in any event.

I stopped along the road at a shop to look at the spectacularly colored pearls on sale. There were gorgeous pinks, greys, silvers, and blues. They looked especially stunning on the rich tan skin of local women. The pearls were totally out of my price range, but I loved looking at them. The saleslady saw my interest and said, "I'd like you to have this shell from which they come." I received a lovely, polished mother-of-pearl shell, which I enjoy to this day. She wouldn't accept any payment. Not everyone is out for the almighty dollar!

Sailing in the Canary Islands

In 1996, Hajo gave me a surprise birthday present: a week working on the crew of the *Winston Churchill*, a tall three-masted schooner usually reserved for training British young people. Once a year, adults could sign up for the crew and sail in the area surrounding the Canary Islands. What a splendid birthday present! A wondrous time at sea! The first night I learned you had to know the direction of the wind if you needed to "feed the fish," or else you'd end up wishing you had checked!

We were forty mates divided into three crew groups on eight-hour shifts, with five professional crew members that included the captain, navigator, and engineer. We each had three opportunities to be at the helm steering the ship, always with a professional behind us in case we made mistakes. I had a particularly challenging turn steering the ship into port in a major gale-force wind. The sails were taken in and the captain called out the directions for me to follow. If I wanted to go right, I had to turn the wheel left, and vice versa, all counterintuitive! I succeeded and was grateful that a professional sailor had my back.

I loved being on the night watch, reporting on the lighthouses as they came into view with their identifying signals. It was beautiful to watch the stars, feel the sway of the ship, and hum to myself and the sea quiet songs of happiness. I was particularly gripped by Columbus's courage to set sail from El Hierro, the westernmost island of his journey, into the unknown on a boat one-third the size of ours! He has been rightly criticized for his actions against the Indigenous people of the lands he found. But he was still one gutsy individual. It was fun to climb the rigging and trim the sails, and to ride into great waves on the bow as porpoises dived back and forth in front of us! So many happy memories for this gal who grew up so far from the sea!

Scuba Diving around the World

Any discussion of circumnavigating the globe would be incomplete without considering the oceans (both above and below), since saltwater covers 70 percent of the surface of planet Earth. One specific method of exploring the depths is scuba diving.

My son Eric, daughter-in-law Lanell, and I all learned to scuba dive at about the same time, but independently. They were training in Puget Sound while I was getting certified in Turkey. I was about sixty years old at the time. Eventually we made many dives together. On our last trip to Cozumel in 2005, my grandson Brendan also earned his diving certificate. What fun to have three generations of us diving together. He was turning somersaults alongside the schools of fish. We watched them and sent our laugh-bubbles skyward through the clear blue waters. We swam in tandem behind our dive master, seeing together the underwater wonders we had only talked about with Brendan before.

I made 170 dives before I gave up the sport. Through scuba, I was introduced to underwater worlds: the Caribbean, Puget Sound, Hawaii, the South Pacific, the Indian Ocean, the Red Sea, and the Azores. I saw more than eighty dive sites. I was delighted by the multicolored brilliance of healthy reefs and dismayed by the devastation of global warming in the graveyards of staghorn coral.

Scuba was addictive. A popular slogan at the time declared: "Life is diving! The rest is just a long surface interval!" I was attracted to scuba when I realized there was a grand underwater realm of peace and beauty. It was a spiritual awakening. We could live life splashing on the surface, or we could dive (and live) deeply and discover a whole other dimension, which completed and gave meaning to our fun on the surface of things. I remember walking on a sandy beach in Hawaii into water up to my waist. I put on goggles and looked into the water. Scores of multicolored fish swam around my legs, and I hadn't even known until I saw it for myself.

During my summer in Israel in 1980, I was able to spend time by the Red Sea. I loved snorkeling and wanted to follow the brightly colored fish over the underwater cliffs. The experience inspired me to learn to scuba dive. Some years later, I joined a group diving in the Red Sea. I was about forty feet down when a gigantic sea turtle slowly ascended toward me from the depths. When it was a few feet from me, it levelled off and looked me in the eyes, and we swam together in rhythm side by side for many minutes. What a magic, mystical time! The turtle gave me one last long gaze and sank back into the depths, leaving a silent but unforgettable memory.

Along the Maldives islands, I saw many beds of blanched coral dying from global warming. I also saw ocean rangers tending turtles whose shells had been sliced by propellers from passing ships. Our group of divers saw a large grouper in a small sea cave. It had caught a puffer fish in its mouth. The puffer fish had puffed up so much that the grouper, its eyes bulging, could neither swallow it nor spit it out. They were locked together in a death struggle. We debated futilely about which, if either, survived.

In the Seychelles, I joined a diving group and made dives around the picturesque islands with their granite boulders and wind-rocked palm trees. We spent a surface interval on one of these islands, enjoying the warm sun on the sandy stretches, the rustle of the palm fronds above us, and the fresh air as we relaxed on the beach. On our return dive to our anchored boat, we were swimming down about forty feet when a large shadow passed over us. I thought it must be a dive boat, but no! It was the elusive whale shark, fifty feet long, the quiet denizen of the deep. Its massive grey form slipped along silently above us. We felt so minuscule in the presence of so large a creature.

I made some of my most rugged dives off the Azores islands. My husband was stationed there during the summer of 1996. We lived on Terceira. I never forgot that I was living on an island in the midst of a very active ocean. We scuba divers would drive around the island looking for a spot to enter the water. Often it was off a seawall with fifteen-foot swells. We would step gingerly off the wall into the swell, which dropped dramatically, holding us in its grip until we descended even deeper, away from shore into an underwater world of volcanic canyons, cliffs, and caves. Moray eels poked their heads out from lava crevasses while red, black, and gold fish, looking like small German flags, darted past. A giant octopus with several partial tentacles seemed to be cowering behind a volcanic boulder, perhaps trying to escape from the deep-sea fishers who had amputated its legs earlier. Even at depth, we were rocked by the ocean currents. We flew past small, bright-blue, iridescent fish.

To resurface, we had to ride the swells up the seawall, grasp some ladder rungs, and hang on as the sea dropped away from us, leaving us hanging with all our gear dangling down around us. With the help of others, we handed up fins, face masks, and air tanks until we could climb up to the top. We always made it, but it was a challenge!

Almost every dive offered us something unique, depending on the dive site. Exotic, brilliant tropical fish swam in splendid schools; rather scary eels stuck out their heads from coral crevices. We saw timid octopi, heard clicking shrimp, and examined tiny incandescent nudibranchs. One dive, I had to enter the water by falling backward from our Zodiac into a school of circling sharks. They were not hungry, and we survived the dive without incident. During another dive, I almost stepped on a giant ray that was lying on the sea bottom. I thought it was a sandbar until I saw its spine. Twice we swam past a scattering of toilets dumped on the ocean floor, apparently from overloaded ships. I was always moved by the silence of our dives, the depths of translucent blue water not unlike cathedral windows, and our silent celebration of partaking in the wonder of the ocean.

Scuba Challenges

Scuba is not a sport for the weak-hearted or the unaware. The diver must keep track of her equipment and gauges, the depth, her partners, available air, and many other factors. Three times while diving I faced pulmonary edema—my lungs filled with liquid, and I lost all strength. I was essentially helpless. The first time, it happened in the cold water of Puget Sound. Fortunately, Eric, who was my diving partner at the time, came to my rescue. We were swimming out together from a beach, using our snorkels to save our tanked air for the dive further out. We were moving slowly out to our dive site when suddenly I could not keep up with Eric. I dropped back and slowed my kick, trying to catch my breath. It didn't work. I still had my flotation device, so I wasn't going to sink, but I couldn't breathe enough to continue.

Eric stopped and asked if I was okay. I signaled I needed help. I couldn't catch my breath. We had completed water safety classes. I knew enough not to panic, and he knew how to swim me to the shore. A few hours later, I was breathing normally again. We didn't know what had happened. Later, diving experts diagnosed that I'd had cold-water-induced pulmonary edema. I stopped diving in cold water. Incidentally, some medical analysts believe that many victims of the Titanic tragedy didn't simply drown but died from pulmonary edema in the freezing sea.

Fifty dives later, in Fiji, my last experience with pulmonary edema ended my participation in scuba diving, and almost my life. My diving partner was Jon Lundman, a scientist and artist. He was the design director for NASA's Golden Records, launched aboard the two Voyager spacecraft in 1977. They present the sights and sounds of Earth on twelve-inch, gold-plated copper disks. The disks provide specifics about our planet, human beings, flora and fauna, and how to retrieve this and other information when found by intelligent beings in outer space.

Jon and I were diving buddies on several dives as part of a cruise in the South Pacific. We had an excellent dive master. However, when we arrived in Fiji, we were required to dive with a local dive master. He did not follow the protocol of descending to the maximum depth early and then gradually working one's way to the surface. He had us dive down to a cave, up to a coral wall, and back down to a swim-through, followed by more ups and downs. By the time we had finished our dive, I felt fatigued, but not distressed. We surfaced right by our dive boat. Jon was nearer than I, so he climbed up the ladder and onto the boat.

I followed, but I suddenly realized I was having a pulmonary edema attack. I could feel that my lungs were congested. I did not even have enough strength to hold on to the ladder. As I let go, I slipped back down into the water. The waves washed over my head. I managed to inflate my flotation device enough to keep my face clear of the water. As I rose and sank with the swells, I began to drift away from the boat. Jon, still in his diving gear, except without fins, saw my dilemma and called for help as he entered the water to help me. No one on the boat came to our aid. Without fins, Jon could not swim with us toward the dive boat. He stayed with me as a strong current carried us ever further away from help.

The sea was rough. We would rise on big waves and drop down in the hollows. Every time we were on the crest of a surge, I could see that we were ever further from the dive boat. Strangely, I thought of Jon's golden disks moving through space as we drifted through the vast ocean. I kept saying to myself, "Don't panic. Keep breathing. Don't panic. Keep breathing." Jon kept checking on me, reassuring me. Visions of being lost at sea slipped in and out of my mind. I remembered a recent story of a pair of scuba divers who simply disappeared. Finally, far in the distance, we saw our own cruise ship dive master waving at us from the dive boat. He had seen us and insisted that the ship's captain maneuver immediately to our rescue.

I felt totally weak and helpless as I was pulled on board. I lay shivering on the deck as people tried to confirm that I was okay. They helped me out of my cold wetsuit and into warm clothes. I felt relieved and grateful to be back on board. Later, on shore, I was able to walk back to our cruise ship. The ship's doctor checked me out. A few hours later, my lungs were completely clear and my strength returned. How humbling, for the independent woman I think I am, to realize that I owe my very life to others: Eric, Jon, and the dive master. Actually, we all owe our lives to so many others, known and unknown! Now, I enjoy fish in the aquarium and snorkeling in safe waters. Besides, there are many other ways to discover and enjoy the water world.

Here's to Small-Boat Adventure

Travel is not only about going and arriving in special places, but about how one navigates to and enjoys the destinations. Take, for example, Hallstatt, a stunning village in Austria beside a quiet alpine lake. Hajo and I arrived there at a quaint guesthouse right on the lake, where we had reserved a room for the night. We both thought it would be an appropriate setting for celebrating his sixtieth birthday. He did not suspect what awaited him.

I had reserved a small rowboat. Instead of going to the usual afternoon coffee hour, I suggested we go rowing on the lake. I told him I really wanted to row on these clear, silent waters. He agreed to go with me if I rowed. He had driven to our destination and wanted to relax. We strolled down to the dock and settled into the boat: he on a bench in the back and I at the oars. I placed my backpack in the bow.

"Why the pack?" he asked. I said that I always take water to drink, rain wraps, and a camera for all eventualities. "You know mountain weather can change quickly." We set out for a quiet, easy row to the center of the lake. It was idyllic! We paused to drift and enjoy the view of the mountains rising above us. A church tower and village houses reflected on the lake.

"Okay," I said. "Now we celebrate your birthday". I opened my pack and pulled out a bottle of champagne, two glasses, a small birthday cake, and a dozen birthday cards from family and friends around the world. There in our simple rowboat, we popped the cork, sipped champagne with many toasts, and enjoyed cake and cards in one of the most scenic settings in Europe! Happy birthday, dear Hajo!

One Week in a Canal Boat

Another boating celebration took place years later in southern France. Our son Eric had looked into renting a canal boat on the Canal du Midi in Provence and invited us along. This time we were celebrating our fiftieth wedding anniversary, and he and Lanell their twenty-fifth. Our grandchildren completed the crew.

Eric piloted our rented canal boat, complete with a kitchen, a lounge, a deck for outside relaxation, and four bedrooms, each with a separate head. For one memorable week, we proceeded slowly along the waterway through picturesque farming villages, vineyards, and shady tree-lined passages, and up and down through numerous locks. Occasionally, we rode the bikes provided on the boat. We cycled along the paths paralleling the canal or explored the surrounding countryside.

What a relaxing, delightful way to celebrate our marriages and enjoy this corner of France. We began the morning with fresh, crisp croissants from local bakeries. We paused in the afternoon for wine tasting at selected wineries. We had time to absorb the atmosphere of this stretch of southern France: the sights, the sun, the fragrances, the flavors of local specialties, and the engineering of this remarkable canal.

Boating in the Pacific Northwest

Over the years, Eric has become a boating enthusiast on the inland waters that run from Olympia in Washington through Puget Sound to Desolation Sound in Canada, east of Vancouver Island. He and Lanell spend much of their vacation time exploring islands and inlets in this spellbinding water world, and they invite me along on many of their voyages on their Navigator powerboat. I have fallen in love with this means of transportation. I have learned that travelling around the world in eighty years need not be only life-threatening adventures, or spine-tingling encounters, or wild discoveries, or unique events. A good dose of relaxation, moving with tides and dropping anchor in quiet coves, is also a fulfilling way to come to know the world.

This is not to say that such boating is always calm. Take our trip through Canadian waters in 2022. Our first major goal was a visit to Princess Louisa Provincial Marine Park. We were always on the lookout for dead heads—logs just under the water that could cripple the boats of unsuspecting navigators. Ferries, which had the right of way, had an annoying habit of suddenly appearing around an island, requiring smaller boaters to make a quick maneuver to avoid contact! We constantly had to be aware of everything around us: a boater approaching from behind at twice our speed, the spray of whales in the vicinity, and our boat wake, which might cause problems for kayakers.

We were motoring up the Jervis Inlet toward Princess Louisa Park when we suddenly saw a tremendous splash far in front of us. We couldn't believe it was a breaching whale, or the like. The clouds hung like white gauze on the forested mountains above us, limiting our vertical vision. We continued along the inlet when a second splash came directly in front of us. We veered to the left, giving it plenty of room.

Then we saw that the splash was caused by a gigantic log dropped from a helicopter. As the log was released, the helicopter shot straight up, freed of its heavy load. We learned later that this was logging Canadian style, from forests too steep for roads. Crews harvested the trees and connected them one by one to metal cables attached to a helicopter, which then flew them to the closest bay. There they were released, collected in tight booms, and later pulled by tugboats to local mills.

Several hours later we reached Princess Louisa Park. Cliffs thousands of feet high rose fjord-style at the end of the inlet. Dozens of waterfalls dropped straight down or cascaded like white cords down the cliff walls into tall, thick, green forests. Somewhere in these forests, the dozens of rivulets combined into one grand river, which plunged a final several hundred feet as a mighty waterfall into the inlet. We tied up to a small dock nearby, just out of the spray from the falls, where we stayed for several days.

We spent sunny hours gazing up at the display and listening to the roar of the grand waterfall, and watched as a two-masted schooner glided into the inlet and dropped anchor, manned by a group of teenage mariners. We hiked several small trails, and were startled on one occasion when a bear's head menacingly appeared around a corner! It turned out to be artificial and was nailed to a tree to scare unsuspecting hikers. We explored the cove in our dinghy, marvelling at the unscathed wilderness around us. Princess Louisa Inlet, in its pristine grandeur, inspires everyone fortunate enough to visit this Canadian treasure.

We motored back out of the inlets and then north for several days to Desolation Sound. As we travelled, the mountains rose ever higher, with snowbanks and glaciers way up above us. We anchored at several provincial marine parks en route. We watched bears on deserted beaches as we motored along. We examined native petroglyphs, one of a red salmon on the side of a smooth cliff at water's edge. At Toba Inlet, waterfalls plunged from rocky heights into a turquoise bay colored by glacial silt. We enjoyed the long golden sunsets reflected in quiet settings.

Eventually we reached Desolation Sound and Von Donop Inlet, where the waters are so far inland that there is little to no horizontal tidal exchange. The sea rises and falls, but does little stirring, allowing the water in the bays to warm in the summer sun. Instead of the water becoming colder the further north we travelled, it became warmer. We spent several days swimming, kayaking, and soaking in these refreshing, comfortable waters. We sat on flotation chairs tethered to our boat, enjoying our very happy hour drinks. Such unexpected fun in the Canadian marine parks.

All too soon we had to motor south again. We stopped to watch whales several times, with their spray and fins appearing alongside our boat. We dropped anchor at Montague Harbor, where I remember the middens: white beaches, not of sand but of generations of clamshells accumulated from First Nation people over the centuries. The presence of tribal inhabitants is often evident in the Pacific Northwest. The Canadian government has even worked with the Inuit people to establish their own territory, Nunavik, northeast of Quebec. This is part of an ongoing effort by Canada to recognize the country's Indigenous people and find ways to help them preserve tribal rights of land and culture. Meanwhile, I have Eric to thank for introducing me to the wonders of small-boat travel.

Figure 23. The Winston Churchill sailing vessel

Figure 24. Working the ropes on deck

Figure 25. The Canal du Midi in southern France

Figure 26. Eric and family navigating the Canal du Midi

Figure 27. Boating in the San Juan Islands, Pacific Northwest

Figure 28. Preparing to Scuba dive

Synchronicity Blows My Mind!

May 18, 1980—Mount Saint Helens explodes in a mighty eruption in Washington state. The same day, the Congress Hall in Berlin, Germany, collapses from unverified causes.

The telephone rings. It's a friend you haven't seen in months but were just thinking about minutes before.

I tell my neighbor I am going to Australia in a few weeks. So is she! We have not mentioned it to each other previously.

Such events—or encounters—happen, and I have given examples in my previous stories. Others, which I relate now, stand alone.

The Telephone Call

My sister Jane once told me about the morning she felt compelled to call a girlfriend, Karen, on the phone. She pushed down the urge because she hadn't talked with her in years. The drive kept haunting her, so she finally found Karen's telephone number in an old address book and dialed. She listened as the phone rang five times, six times, ten times. "Okay. I'll hang up," she said to herself. Then she heard the familiar voice: "Hello."

"Hi, Karen. How goes it?" Jane responded. Silence. Then Jane heard Karen crying on the line. When Karen could talk, she admitted, "I was just starting to take a bunch of pills, when the phone rang. I've been having a tough time and I can't take it anymore. I have to end it." Jane's call had interrupted a suicide attempt.

Jane listened and talked with her for over an hour. She eventually encouraged Karen to find the help she needed. But Jane always wondered what had caused her to call Karen at this life-threatening moment.

Snakes

When my family was living in rural Montana, Mark, 16 years old at the time, cautioned me on a warm, sunny day as I was about to leave on a jog: "Mom, don't run along the dike. You'll run into a rattlesnake!"

"Hey, Mark, I'll be okay. I run along the dike all the time and I've never seen a snake."

"I'm just telling you, Mom."

Well, I set off running along the dike, chanting a silent mantra—"Snake, snake, snake"—as I focused on the tall grass beside the trail. Halfway across the dike, I heard behind me the telltale rattle of a prairie rattler. There, a few inches from where I'd stepped, rose the poised head of a coiled snake. I quickly backed off and it didn't strike. This was the first time Mark had ever warned me about snakes. Coincidence?

Twenty years later, Mark visited me in Boulder, Colorado. We were preparing to hike in the hills north of town when Mark announced, "Mom, we're going to see a rattlesnake today."

"Mark, I have gone on hundreds of hikes around here and I've never seen a rattlesnake." But, sure enough, that day we had to walk around a baby rattler: just one button on its tail, but for sure a rattlesnake, right on our trail.

Frau Hanna Jolly

In the 1970s, Hajo and I were living with our two sons in Havre. Hajo was a professor at Northern Montana College when he realized that one of his students was from Berlin, where he was born and spent his early years. Hanna Jolly was the wife of a US Air Force officer stationed near Havre. We met with them socially for several years, but lost contact when they left Havre on a new assignment.

Six years later, both Hajo and I had sabbatical leave from our teaching jobs and decided to spend the year in Berlin with our sons. The first week in Berlin, we visited a remote museum on the west side of the Berlin Wall. There, in a visitor's register, we found her name. Hanna Jolly had visited this same museum the day before us.

Two weeks later, Hajo was walking down a major street in the Berlin Center when he saw Frau Jolly walking toward him . . . one of how many thousand Berliners!? Another time, I met her quite by chance on a city bus. It was like we were destined to stay in touch—or was it coincidence?

Paris

I love all things French, and in 1961 I was delighted to receive a Fulbright teaching fellowship to a lycée in Fontainebleau after graduating from the University of Colorado. One day, I decided to visit Paris and took myself on a tour along the Seine. I was crossing the Pont Neuf when a very insistent Frenchman called after me, "Mademoiselle, mademoiselle!" I ignored him. The French were known to accost young ladies. I walked more quickly, determined to leave him behind. But he caught up to me. I turned around. There behind me was my French professor from three years earlier at CU. He had recognized me but forgotten my name. What forces had come together that we were both crossing the same bridge at the same time?

Animal Tales

During the summer of 2019, my oldest son Eric and his wife, Lanell, invited me on their boat for a few days of cruising in the San Juan Islands north of Seattle. I had kept it secret that I hoped to see some orcas, the local whales, but after several orca-less days I told Eric about my wishes. I added, "Eric, I would be happy just to see a walrus instead."

He replied, "Mom, you have as much chance of seeing giraffes up here as seeing a walrus." We laughed and then forgot about our conversation.

A couple months later, I was walking around neighborhoods in Boulder that I had not visited before. I was surprised to come upon a house with a majestic walrus statue in the front garden! The owner, who was working in his yard, explained that he had purchased it from an artist on a trip to Mexico. I told him how much I liked it!

A few weeks after that, I was hiking above Boulder when I saw the same man on a trail with his wife, whom I had met some time before in a discussion group. I told her I really liked her walrus. She smiled and added, "If you like that walrus, you should see the giraffes in the backyard."

"What?!" I exclaimed, laughing. Then I told her about the walrus-and-giraffe conversation from my boat trip with Eric. I returned to their home to document in photos their remarkable statues. What are the chances of discovering two such disparate animals, both in a boat conversation in Washington state and in a garden in Boulder, Colorado?

Once upon a Hike

When I first returned from Europe to our home in Boulder, I told my husband I was itching to get out and hike up a mountain I had heard about. I knew nothing about it except that the tundra flowers were especially nice. He was busy, so I took off alone. It really wasn't a mountain so much as a series of high spots on a rocky ridge, above which rose South Arapahoe, a major summit in the Indian Peaks area. I wandered around in the high tundra area, trying not to trample the high-country vegetation (which was lovely) as I explored. There were many piles of rocks scattered over the hillside. Off in the distance, I saw one that caught my eye. It was getting late, and I thought I should be getting down rather than going up further, but these particular rocks seemed to be calling to me.

I rationalized it would only take me a few minutes more and be a good destination. So, against my better judgment, I headed up the hill to these boulders. Since I was there, I might as well climb up to the top. It was a short, easy scramble, but what I discovered on the flat, topmost rock was all so unexpected, so strange, that I cannot explain it to this day. There on this basically nondescript rock was a huge, very heavy typewriter just sitting there. No paper. No typewriter ribbon. No explanation. Who had carried it all the way up there? Why? What had made me want to go to that rock among so many? These were questions I pondered as I hurried down to beat nightfall.

An Incongruity

I have always enjoyed volunteering with various organizations, doing trail work or killing invasive species. On one trip to the Arizona desert, my group was out to rid the landscape of a particular plant that was taking away water from the native species. We were divided into pairs: one person with a huge bag to collect the guilty invaders, the other seeking them out. My partner was busy pulling out the culprits in a small area while I walked along our assigned section to find more of them.

We were about a half a mile away from any roads when I spotted a shiny pastel object in the distance. I walked over to investigate and found a porcelain statue of an angel standing all alone in the sand. It was about two feet high, lovely and serene in this most inhospitable dry wilderness. I called my partner over. She was as mystified as I as to why this angel was in this spot, which was just flat desert like every other spot. Someone had placed the angel there for some reason. It seemed it was a sacred site. We stood there in silence for a while, then continued our mission to free our corner of the world from evil weeds.

What's Going On?

I am convinced there are forces in the universe that we cannot really understand, prove, or disprove. Some people call them God and see divine intervention or Karma, while others sense nothing at all. Things happen. Lives intertwine. Personally, I am left with feelings of awe, wonder, humor, sadness, delight, bitterness, gratitude—the whole gamut—as I relate these stories: some mundane, others gripping, and many others deeply moving, perhaps like your own life stories.

Figure 29. The garden walrus

Figure 30. The back yard giraffes

Figure 31. Discovering an angel in the desert

Retirement—Not Really!

When Hajo took final retirement, we returned to Boulder, where our long-rented duplex was free for us to move into our retirement mode. I had expected to settle down, but adventure called, and I answered.

Haiti as I Lived It

I became involved in Haiti because I thought that, knowing French, I might be of some help in what I thought was a French-speaking country. Actually, the French colonists had imported slaves from Africa to work on their sugar plantations. They had taught them French, which became combined with African languages, so that Haitian Creole is the real daily language of Haiti. However, the school system is supposed to run in French, and official business and documents are in French. Haitians were happy to communicate with me in French since few of them knew English at the time.

My first contact with Haiti involved sponsoring a Haitian girl through Compassion International, a Christian mission program, in an unknown destination in rural Haiti. Aramanthe was a teenager in second grade. I wrote the sponsoring program and asked why she was only in second grade. They explained that she often missed school because of toothache. (Actually, thousands of Haitian children suffer from toothaches because raw sugarcane, which is a common part of their diet, causes their teeth to rot.) Compassion International suggested that the best way to help Aramanthe would be to help her purchase a sewing machine. She could take sewing classes in the school and would then have a way to make a living.

I purchased one for her. For several years, I received photos of her at her sewing machine and thank-you notes written in Creole. Then I heard no more. I wrote the project. Aramanthe was no longer in the school, and they couldn't learn anything about her. She and her sewing machine had dropped out of sight. I have always wondered what happened to her. What is an appropriate way to combat poverty? Was donating anything to Haiti like pouring money into a black hole?

In 1998, I heard of the Colorado Haiti Project, a mission project of the Episcopal church in Colorado. CHP was seeking volunteers to travel to Haiti to help with a medical mission and see for themselves the conditions there. I decided to go. As the only French speaker on the mission trip, I became an interpreter. Many Haitians sought me out to tell me their stories. This was the first of six mission trips I made to Haiti and Petit-Trou-de-Nippes over the next fifteen years.

First Impressions of Haiti

Ten of us flew from Miami into the capital city, Port-au-Prince, with fifty duffel bags of supplies. At the airport, the heat and humidity were stifling. Clearing customs took several hours. An open-backed truck, known as a camion, waited for us. Police guarded the area around us as we loaded the bags. We sat in the back on hard benches with half a dozen Haitians, seeking rides to our destination. Then began our perilous seven-hour trip to the mission site, Petit-Trou-de-Nippes, on the southwest peninsula.

Our camion roared through the capital city, past the gleaming white government palace and the independence statue of a native Haitian blowing his conch shell. We passed slums, where pigs foraged for food in the garbage strewn out to the ocean inlets and people picked their way through the filth. We avoided several intersections where groups of Haitians were burning tires to protest some issue. Around us, brightly painted and overloaded taptap buses lumbered along with their signature names: "Praise to God," "Hallelujah," "Merci Jesu," et cetera. Eventually, we were out of the city and stopped for fuel at the Filling Station of the Immaculate Conception.

We continued on badly rutted roads through small villages with tiny multicolored shops. Children ran along beside our camion, calling out, "Blancs, blancs, blancs," as we white folks drove past. We rocked through the rough streets of several larger towns until we reached the Grande Riviere. We had to wade across this relatively shallow, broad river.

We waited on top of the steep bank on the other side, where our camion would pick us up. As we waited for our transportation, another truck approached the river, stopped, geared down, gunned its engine, plunged into the river, sprayed walls of water on both sides, and then struggled up the slick bank to the flats where the road continued. Partway up, the truck's tires started spinning in the mud. Scores of shoeless Black men with bare backs and torn shorts raced over to help. They clawed up dry dirt with their bare hands to throw under the spinning tires, giving them traction. As the rear of the truck jolted and slid from side to side, the men frantically dug and threw more dirt, willing the truck up the bank. I watched, horrified at seeing human beings used so inhumanely, the skidding tires narrowly missing them. The men only stopped when the truck finally made it over the lip of the steep bank. Then they stumbled over to the shade of some trees and collapsed.

Our camion was next. It blasted through the river and up the bank under its own power. We all cheered and climbed back up on our hard seats. Two rough hours later, we made it to our destination.

It was dark as we drove into the mission compound. We could hear the roar of a generator in the background as we unloaded the camion into the partially built church, which was lit by a single light bulb in the sanctuary. We were road weary, but managed to pitch our tents in a circle in a field illuminated by the truck's headlights. It was still hot and humid as we crowded into a small dining room around a long, rough table. Under another single bare light bulb, we bowed our heads in thanksgiving for having made it safely to the mission—no small accomplishment.

Over a supper of rice and beans with some mouthwatering, tangy spices, we met the mission leaders. They told us we would hear a wake-up call the next morning. Then we took turns showering in a welcome cold-water shower installed against the back of a shed. We crawled into our tents, and we fell asleep hearing drums and songs from what we learned the next day was a nearby voodoo center. Welcome to Haiti!

The Mission

The next morning, we were awakened at sunrise by the crowing of a rooster strutting proudly in the center of our tent encampment... much too early, but everything worked according to natural light. The mission generator was a rare source of artificial light. It seemed like hundreds of Haitians were already waiting in line to see the mission doctors and nurses. We quickly set up a pediatric clinic in the church sanctuary. Half of the large altar was a dispensary stocked with drugs against pain and worms, antibiotics, Band-aids, and vitamins. The other half was an examination and treatment table for our mission pediatrician, Dr. Eben Carsey.

I remember holding a little girl's head firmly as Dr. Eben searched for and extracted a partially sprouted bean from her little nose. She remained stoically silent throughout the trying procedure. Next, we administered soothing ointment on a pathetically thin boy's back, which was covered with dime-sized sores. For me, it was a humbling experience to work at an altar where spiritual and physical compassion found expression.

By then, the children were arriving for classes. The school consisted of a framework of poles set upright in the dirt, with crossbeams nailed in place to support a series of tarps that provided shade for the desks. These were benches with rough writing surfaces. Each class was separated by rickety blackboards from which the lessons were presented. The students had no books. Some children had thin notebooks into which they copied notes from the blackboards. The other students sat and listened. Many of the lessons were learned by rote memorization. The students ranged in age from three to five in the preschool classes, and six to twenty-five in the other classes. Adults who had not had a chance at education joined children in these classes so they could become literate. The children themselves loved being together in school and sang in unison to begin their days.

The six teachers were totally dedicated to their students and worked hard to teach them the fundamentals of math, French, and Haitian history. Only one teacher had more than a sixth-grade education. He taught sixth grade until he died of a stroke on the way to school a few years later. The stately Madame Odette was the school coordinator, and also a teacher. She maintained discipline with a switch and a smile. She knew each child and each family and was responsible for keeping the school going. I worked closely with her over the years and have nothing but respect for her and her many contributions to the school and church.

She asked me if I would return and give teacher seminars, especially on science. She admitted her own ignorance. She and the teachers were amazed to learn that men had gone into space and walked on the moon, that there were other planets, that dinosaurs had existed, that there were maps. In short, this was a third-world community trying to catch up with the rest of the world. The progress over the next fifteen years was remarkable!

Highlights and Disasters

In succeeding years, I returned to Haiti, giving teacher seminars and working with students and the community in general. Every trip had its tragedies and successes. On one mission, we arrived in a mostly empty town. Everyone was gathered in the Catholic church for the funeral of a teenage girl. She had died of a botched abortion. Great cries of grief emanated from the church as we passed by. Since then, the country has focused attention on safe sex, contraception, and organizations supporting teenage girls.

I taught many science seminars, which teachers from neighboring towns often attended. I remember after one astronomy session, a visiting teacher thanked me profusely for the seminar, saying he had learned so much. He added that as I spoke, he could "see that all those stars and galaxies were glimmers in the eyes of God." His poetic vision gave a perspective beyond the mere facts I had shared. In some science sessions, a young Haitian student would sit in the back. In the evening, he would seek me out and ask questions about science and its applications. He has since graduated from high school, taken special training, and become the community plumber. The general science courses in the school were modified to include specialized agriculture programs that benefit the entire community.

In and around my time spent giving the seminars, I became involved in community affairs. A farmer in an isolated village about four hours walk from Petit Trou de Nippes had donated a piece of land for a school. Several young women were trying to start one on this land and asked if they could attend my seminars. I said they were welcome. They arrived in light dresses and sandal-type shoes after the four-hour walk. I doubt they had had anything to eat since the previous night. We found some fresh water for them and a few nut bars.

When they arrived, we were making flash cards for a simple grammar game in French. We shared our materials with them, but they had never cut with scissors or written with markers. They spoke only Creole. They were keen to learn and to become real teachers for the children in their village. Such motivation was overwhelmingly touching. I wished it was enough to bring change, but motivation alone cannot combat the devastating forces of poverty, earthquakes, epidemics, tropical storms, and political instability that overwhelm Haiti every time it begins to make progress.

Life in Haiti can be brutal. On one trip, our mission doctor was called to the police station in Peti Trou de Nippes to examine a man who had beheaded his sister with a machete in a family argument. I went along to translate. He was hoping to plead self-defense because she had attacked him with a broken bottle and he had gashes all over his arms. There he was, a slight, broken man, eyes turned down, pathetic. Our doctor was to examine his wounds as evidence in his defense. We submitted his report, but it was later rejected because I was not an official police interpreter. We never learned what happened in this tragic case.

Over the years, I became acquainted with many Haitians. Yvon was a carpenter who was proud of his work and discussed it with me. He made houses, headboards for beds, decorative fish statues, and coffins, which were in great demand. He cared well for his family and wanted a good education for his children. Every year he became more crippled until he admitted to me that he didn't know how he would be able to care for his family. He could no longer saw boards due to his infirmity, and the doctors couldn't help him. An electric saw would be helpful, but required a generator and fuel, which were difficult to obtain.

I listened sympathetically and he thanked me for listening. I wished him well, but he looked depressed as I left. Three years later, I returned to find a smiling Yvon working in his new business. He ran a cell-phone recharging station powered by solar panels. Suddenly, despite general poverty, everyone had a cell phone but no one had electricity. For a small fee, he kept the phones running and made a living for his family. This was Haitian ingenuity.

Jean-Pierre was a very bright teenager, an orphan who had been adopted by a Pentecostal church in the neighborhood. He attended the public school but wanted to sit in on my seminars. He also wanted to improve his French, so he talked to me about his life every time I went to Haiti. He hoped to become a Pentecostal preacher and church leader. About the time he graduated from high school, he came to me in tears. He had a girlfriend; she was pregnant, and he had been expelled from the church for making her pregnant. She had been denounced by her family. He was repentant for his actions, but what could he do to help her? It was a typical poverty story. Over the years, she developed tuberculosis, and then their baby did also. Jean-Pierre worked at odd jobs wherever he could, trying to feed them and obtain medicine. Later, I was told that both his wife and baby had died and he had left the area. (Such poverty and hopelessness contribute to the violent, lawless Haitian gangs.)

One particularly dramatic episode occurred when we were returning from our mission at Petit-Trou-de-Nippes. We drove for miles on the roughly rutted road in our camion behind a large truck overloaded with charcoal sacks headed for market in Port-au-Prince. As we neared the Grande Riviere, the charcoal truck lurched to an abrupt stop. I watched as a large Black woman slid silently and uncontrolled from her perch on top of the sacks and crashed brutally to the road below. She did not cry out, but she lay there in the dirt with a broken bone jutting out from her lower leg, which was covered with charcoal dust, dirt, and blood.

Our mission group immediately climbed down from our camion. Our doctor quickly applied a tourniquet. The bleeding stopped. We collected all our drinking water and washed the wound. We splinted her leg with cardboard cartons from our camion. Then we realized she was alone. No one from her truck acknowledged knowing her. She herself said nothing. We knew that before she could be taken to a hospital, she would need a sponsor. It took us an hour to find a man willing to transport her, pay the required minimal charges, and arrange for food for her hospital stay. We took up a collection and gave it to the volunteer. Then this diminutive, wiry man stood while others hoisted the heavyset woman onto his back.

On our last view of her, she was being carried across the river. Her leg in the crude cast stuck out obliquely in front of her. Her torn clothes hung loosely around her, more like rags than coverings. Throughout the ordeal, she remained stoically silent. Two men helped balance her on her rescuer as they waded through the brown river, the water up to their thighs. All human dignity was swept away downstream, but for the moment she was safe and cared for. We hurriedly waded across the river and returned to our camion. A few hours later, we were flying to Miami and a whole other world.

Some Americans ask me about voodoo. Many Haitians practice both Christianity and voodoo, though most pastors tell them they don't need the voodoo charms they are wearing. I once talked with a Haitian man in African garb, covered with gold bracelets and necklaces and a huge gold cross. He was my seatmate on a flight to Haiti and, he told me, a pop singer in Port-au-Prince. He explained he was Christian, but he believed that voodoo was good for many Haitians. He had dedicated his most recent album to a particular voodoo god who had helped his grandfather through many difficulties. In Haitian art galleries, many of the paintings depict voodoo gods that parallel Christian saints.

These are but a sample of the people and experiences that make up Haiti. There are many different Haitis. The gang warfare and violence of Port-au-Prince are not the same as life in the rural farming communities. Despite hardship and poverty,

Haitians are resilient and optimistic. The children love to sing, dance, and laugh together. When the international seed banks deliver supplies to rural areas, the farmers gather together and work out the distribution with their leaders. The American mission folks have learned not to donate physical items to Haiti, but rather funds to help the local communities develop programs and facilities appropriate for their needs. I have seen this kind of aid bring real change for the better in Haiti.

This story does not stop here. The Colorado Haiti Project has been expanded to an entity called Locally Haiti to reflect local leadership in a total community program. Locally Haiti is working with the national health ministry to build a hospital for the twenty-six thousand people in the Petit Trou de Nippes area. When completed, it will be the only one in the region. The mission school continues, with over five hundred students in preschool through high school, trained teachers, and agricultural and girls' leadership programs. There are still many challenges in this area, but I believe the Locally Haiti project is a positive example of development in emerging third-world countries.

Highlights and Disasters | 209

Figure 32. Teaching Haitian teachers a science seminar

Figure 33. School kitchen for hot lunch for 400 students

Figure 34. Before school water collecting chores

Figure 35. Yvon, a local Haitian carpenter with his creation

Cuba Inside and Out

Over the years, I have had a variety of experiences both inside and out of Cuba. During the Cuban Missile Crisis of October 1962, Hajo and I were in eastern France visiting World War I war sites. Hundreds of memorials to the fallen soldiers of many countries dotted the landscape. We were listening to the American Forces radio station when the news came on that the Soviet Union was establishing a nuclear missile base in Cuba, ninety miles from the Florida coast. President Kennedy was establishing a naval blockade around Cuba, fearing that our national security was about to be breached. Hajo and I were thinking war was imminent as we passed through the still-visible devastation of World War I. Could mankind not learn to live in peace?! Fortunately, the standoff lasted only two weeks before Khrushchev removed the missiles in return for the US promise not to invade Cuba.

Almost fifty years later, Hajo and I were on a mission trip to Haiti, where we met several doctors who had been sent from Cuba to meet the health needs of their Caribbean neighbors. They were working in rural clinics to bring care to areas that had virtually no medical facilities. I spoke with one MD, a young woman. Castro had sent her on a two-year assignment to repay her medical education. She had no choice but to leave her husband and infant girl for two years. Her room and board were paid, but she received no salary and had no way to return to Cuba to visit her family, though it was not far to get there.

She realized she owed this service for her degree but was heartbroken to be without her family. Over the years, I have learned that Cuba continues to send their medical professionals to work around the world in areas where they are needed. Many of them are happy to practice medicine; for example, in parts of Italy where doctors are in short supply.

After one mission trip to Haiti, Hajo and I were granted permission to visit Cuba on a special program. The government of Cuba had recently decided to create a standardized system of education. Rural teachers had complained that their students lacked the opportunities of students in Havana and other cities. To equalize education, the Cuban state dictated that all classes in all schools would follow the same curriculum and class activities as outlined and broadcast on television at the same hour. The teachers were required to follow the television directives for each subject. The teachers with whom I talked were disgusted at this system and knew they were not educating the students, who were bored with the whole thing. This program was abandoned the next year and I have no way of knowing what replaced it.

For a few months in the same year, Cuba announced an intention to foster cultural connections with Europe. For example, German and Italian classical musicians were invited to perform in Havana. The classical orchestra conductor Claudio Abbado was invited to conduct an orchestra of professional Cuban musicians and orchestra students. They were excited to be able to perform under the baton of such an outstanding conductor. Then, suddenly, Castro declared that the performance was cancelled. "Okay," said Maestro Abbado, "I will conduct a rehearsal that afternoon. I have my ticket to Cuba and am on my way!"

I was present at the rehearsal. Claudio Abbado appeared in a bright-pink sweater. The performers and audience applauded as he mounted the podium. For the next two hours, we were part of a rehearsal like no other. The students had practiced the pieces so carefully that they were prepared to follow his every nuance. The communication between conductor and performers was intense and inspiring. Abbado drew out music from the performers that thrilled everyone. At the end, performers, audience, and conductor were all standing in a mutual ovation. I was moved by the power of music beyond what any political system could dictate.

Japan in Person

Just when we thought Hajo's teaching days were over, he was called to Japan to teach on various Air Force bases, including on the island of Okinawa. I accompanied him on many of these duty tours and did my share of teaching English as a foreign language in Japanese schools. We had several occasions to meet again with Chikako, my foreign-student sister from my Boulder days back in 1956. She and her husband were both medical doctors in Nagoya. We had the ironic pleasure as Americans of showing them around Okinawa, which they had never visited. They introduced us to the remarkable sites around Kyoto.

Several unrelated experiences stand out from these months in Japan. In Okinawa I met US servicemen who were sent there for a recuperation break from Thailand. They had been given the grisly task of taking care of the bodies of thousands of victims of the earthquake and tsunamis of December 24, 2004. The earthquake in the Indian Ocean, one of the strongest on record, killed over two hundred thousand people. The servicemen were sitting in a bar in Okinawa describing their ordeal, obviously traumatized by the sheer number of victims. I had always thought of soldiers having to confront battlefield realities. Here they were in peacetime with the greatest challenge of their lives. When we say to people in uniform, "Thank you for your service," we can't fully know what this may mean!

Hajo and I often went for meals in Japanese restaurants. We were incredulous about a common offering: Spam sushi! This was a leftover from World War II when Spam was the only meat available.

One of the military bases on Honshu sponsored a climb up Mount Fuji. I was eager to make this climb since my parents had told me about theirs many years before. I carried a wooden walking stick, which received wood-burned stamps at each level of the climb. The trail snaked up many switchbacks, with small shrines along the way. I wasn't expecting what I found at the summit: a whole town with shops, temples, and restaurants spreading out in all directions.

As I walked by the post office, the postmaster took my arm and escorted me inside, bowing to me repeatedly. An employee who spoke English explained it was the opening day of the post office for the season. They would be honored if I would buy a postcard and send it. They wanted a foreigner as their first customer. The action would be recorded on Japanese television. This wouldn't be the only time I appeared on Asian television! Months later, I received the postcard from the summit of Fuji. I had sent it to myself!

The Episcopal Bishop of Okinawa

My husband and I spent several teaching tours on Kadena Air Base on Okinawa. Everywhere you go on this Japanese island you are reminded of the war: the tunnels, the memorials, the rebuilding in which jumbles of structures are compressed together without plans, just wherever there seemed to be room.

We heard there was an English-speaking Episcopal church off base. One Sunday we drove through a maze of streets and found it. It was a small chapel, with a large parking lot. As we were getting out of the car, I saw an elderly Japanese man in a purple shirt and a clerical collar. He was standing beside a car and helping an aged woman beside him. They walked slowly together into the church. I was curious. Wasn't this a bishop? This must be a special service.

It was a typical Anglican service, with liturgy, a sermon, hymns, and Communion. The man in the purple shirt was a retired bishop and had no special duties. He and his wife attended services there each week. They spoke limited English but were regular worshippers. It was several weeks before I could piece together his story, which everyone in the church seemed to know by heart.

As a young man in mainland Japan, he was swept up in the Emperor-worship of World War II. He was chosen to become a kamikaze pilot, and he willingly prepared to give his life for the Emperor on a suicide mission. On the day of his assigned flight, he was in his aircraft waiting to take off when the mission was cancelled due to inclement weather. This was near the end of the war—so near, in fact, that Japan surrendered before this young kamikaze pilot could do his duty.

I imagined he was relieved not to have flown his mission. I was mistaken. There was so much honor and tradition woven into giving one's life for the Emperor that he was deeply disappointed and the honor of his family was at stake. For several years he fought depression, seeking help in Buddhism, oriental philosophy, and meditation. One day someone invited him to a Christian church. For the first time, he heard of the life and sacrifice of Jesus on the cross. He was particularly struck by the meaning of God giving his life for others, the opposite of the Emperor expecting others to die for him. This young man found hope in this new perspective.

He studied the Bible and converted to Christianity. Later he became a priest in the Episcopal church and then Bishop of Okinawa. He had discovered the dichotomy between dying for the Emperor and living as a Christian.

Rescue in the Grand Canyon

Whenever Hajo and I returned to our Boulder home after sessions of teaching abroad, we would volunteer for various causes. Sometimes we worked alone, sometimes together. I chose a trail repair project on the north rim of the Grand Canyon just outside the national park. Hajo left me with a group of twelve and planned to pick me up at the end of the project.

For a week we worked hard restoring a trail from the rim to the river, the Colorado, which had cut the canyon. The trail was overgrown with cacti, scrub brush, and grasses, and rough with loose rock and rubble. It took work with numerous tools, elbow grease, and energy to restore it to a walkable hiking trail, but by Friday afternoon we had completed the work all the way down. We had Friday evening and Saturday to relax and enjoy our accomplishments.

We all gathered near the river in a box canyon to plan our activities. One of our leaders knew of a secret passage that held promise as a group adventure. He asked us to wait as he scouted out an entryway. He was gone hardly five minutes when we heard a scraping sound and a heavy thud like someone falling. We all rushed over in the direction of the sounds. There was our leader, lying motionless on his back on a jumble of rocks. As we stood there imagining him dead, he stirred a bit and moaned. At least he was alive. Three of our group had just completed a wilderness rescue course; they immediately began checking him out and exploring possible options.

After an hour, they determined that his legs were injured and he could not get out under his own power. He might have a concussion, and his back and arms ached. In short, he would need to be rescued. Part of the group would stay with him while others returned to the main camp and brought back tents, sleeping bags and pads, water, food, and meds to set up a medical camp. Two others set off to find a high spot from which to call for a rescue helicopter. The rest would return to the main camp, which would serve as a center of communication. Night fell before a rescue could be realized. We fell into a restless sleep in our two camps, wondering what the next day would bring.

Saturday morning, all but two of us grabbed breakfast and headed to the medical camp. I was left in the main camp with a partner to coordinate communications since we had heard nothing about a helicopter rescue. About thirty minutes later, we saw a man on horseback coming down the trail and went out to greet him. He had a silver star on his leather jacket, a huge cowboy hat, a revolver in a holster on his belt, a small canteen on his saddle, and cowboy boots stuck in his stirrups.

We greeted him, and he called out as he came forward and stopped. "Ah'm the sherf! Ah hear'd there's an injured victim. Ah'm not callin' in no copter till Ah sees fur myself he cain't get out hisself!"

I broke in saying, "This is a very rough remote area where I doubt your horse will be able to go."

"Ma horse goes anywhere Ah go. Ah'm from these parts. Ah knows what ma horse can do!"

"Do you have any water beside your canteen? It's hot and dry down here," I added.

"Ah know what it's like. Ah live here. Now, where's the victim?"

We told him how to find the hidden canyon and medical camp. He left on his horse.

A couple hours later we heard an approaching helicopter. It circled around our camp and the pilot waved. He circled some more and then landed in an open spot we had indicated. He couldn't find the canyon to rescue the injured man. We told him the key signs to locate the medical camp. His copter rose and headed to the canyon. Sometime later, we saw it flying up and away. Our leader had been rescued.

An hour later, the rest of the group returned to the main camp. I asked if they had seen the sheriff. Yes, they had. He had agreed that a helicopter was indeed needed. They had managed to get out a rescue message.

Meanwhile, the sheriff had gone to look for his horse. It had been hobbled, but it broke its hobble and wandered off somewhere. The last they saw of the sheriff, he hadn't found his horse and had run out of water. They filled up his canteen.

We continued to get the camp back in order. Eventually, we saw the sheriff down on the trail that led away from our camp, heading to the rim. His horse was not with him, and he was limping along in his cowboy boots with his hat crammed down on his head. He ignored us. But, with binoculars, we followed his painful process up the trail toward the canyon rim. We saw him stop and sit on a big rock by the side of the trail about two-thirds of the way up. Then we heard it . . . a helicopter coming to rescue the stranded sheriff!

Our leader was not critically injured, and the following year he was back leading trail work. As for the sheriff, we never knew what happened to him or his horse.

Mount Rainier

One mountain dominates the Pacific Northwest in Washington state. You see it as you fly into Seattle. It is the backdrop for photos taken of Seattle and of the islands, parks, and ferries around the city. It is Mount Rainier, a monster of a volcano rising from near sea level to more than fourteen thousand feet, covered with scores of glaciers and volcanic ridges. It towers over all the other mountains in the state, which form the northern chain of volcanoes near the Pacific coast. Further north, Mount Baker seems dramatic until you see it in comparison with Mount Rainier. I remember Mount Saint Helens as more serene and graceful until it exploded in 1980, leaving a grey ruin of its former self. Mount Adams is like a little volcanic brother to Rainier.

I was excited when Pop suggested we climb Mount Rainier when we were on our family camping trip in the state in 1952. We had just finished climbing the Tetons and were optimistic about the possibility of climbing Rainier. I was really psyched, as they say. We camped at the foot of a major route one evening, setting off very early the next morning with crampons, ice axes, and all the gear we would need.

The weather was perfect. Too perfect. As we approached the mountain, we could hear avalanches of rocks and ice cascading down as the warm weather, even at night, melted the normally frozen glacial ice. We watched the sun rise pink on the glaciers. Pop said that it was much too dangerous for us to climb up the route that the avalanches were crashing down. I remember feeling devastated. I'd led a charmed life up until then. I was spoiled! I didn't know what disappointment felt like. It was time I learned!

Eventually I got over the disappointment of not climbing Rainier. I finished high school, went to college, married, had two sons, and didn't return to Washington state for many years. When I did, it was for the university graduation of our son Eric, who had settled in Seattle. Then our second son, Mark, relocated to Seattle and decided to climb Mr. Rainier as a fundraiser for the American Lung Association.

He solicited funds and climbed Rainier for several years before he said, "Mom, you are in good shape. Come climb the mountain with me for the Lung Association!" I was in my early sixties at the time and hadn't even thought of climbing a major mountain. "We climb with guides, and they equip us and give us glacial climbing lessons and take us up the mountain. The hardest part is to get the funds for the Lung Association. Come on, you can do it!"

That was how, fifty years after my first big disappointment on Rainier, I found myself climbing it with my son Mark. Pop was still living and cheered us on from his garden near Seattle.

I was looking forward to the climb! I had a pack decorated with ribbons, one for each Lung Association donation from family members and friends, which made it a community effort! Mark and I hiked up with the rest of our climbing party to a shelter for the first night. It was a lodge with platforms where a dozen of us lay in our sleeping bags side by side. We were told to hydrate generously to be ready for the climb the next day. That meant that all night, climbers were going in and out of a squeaky door to the nearby restroom. No one could sleep with all the comings and goings.

About the time we finally dozed off, someone called out, "Time to go!" It was just past midnight as we ate a quick breakfast and filled our packs with snacks, water, and wraps. We fastened our crampons on our climbing boots. Crampons are necessary for climbing glaciers to ensure a firm grip on the ice. Our climb was on both glaciers and volcanic rocks and sand. To avoid having to stop and take off and put on the crampons, we had to wear them for the entire climb to the summit. We stumbled, tripped, and griped our way up the rough volcanic stretches. We were glad to finally have them on the glaciers, where we roped up. It was fun to be on the rope with my son, climbing together. It was a spectacular feeling to climb across the glaciers with views way down the mountain and across to other summits below us.

We experienced a startling highlight as we crested over a glacier near the summit. We found our way around crevasses from which volcanic steam was rising. It was not visible from a distance, but we were right there in the steam! I felt dramatically vulnerable knowing we were on a very live volcano. Scientists were always saying that Rainier was not extinct, but when you stand by the vapor, you see and feel that Rainier is very much an active giant!

The summit covers a vast crater filled with ice and snow. We walked partway around the crater to the highest point, Columbia Crest, where I signed the summit register. It was a fulfilling feeling, having completed a fifty-year-delayed summit! Far off to the west we could see the tiny skyscrapers of downtown Seattle. It was a lovely clear day to be on the mountain! However, just as warm weather had cancelled my first attempt on Rainier, our guides hurried us off the summit and down the mountain, saying that as the day warmed, avalanches were more likely to develop.

The trip down went much faster than the climb up. After we had descended the steep parts at the top, we could slide down great stretches of snowbanks like crazy kids. Eventually, we reached the trail leading down to the base. It was really warm as we trudged on down, and we were tired when we reached a stream. Surprise! My brother Mike was waiting for us with cans of ice-cold beer! Celebratory brews. Wow, beer never tasted so refreshing!

Some ten years later, Mike and his daughter climbed Rainier, and Mark and I shared with them another such celebration!

Around the World of Biking

I have always admired bikers who have pedalled many, many miles: from Alaska to Patagonia, or Maine to California, or even around the world. I enjoyed following the Tour de France on television or occasionally in person while in France. I was amazed by the endurance of these athletes, the beauty of their streamlined muscles, and their ability to ascend steep mountain passes and arrive on top scarcely winded, then fly down curving roads like birds in flight.

My own biking experiences were much tamer, but nonetheless challenging for me. The San Juan Huts System places small cabins at daylong intervals from each other across the Rocky Mountains, starting from both Telluride and Durango in Colorado and ending in Moab, Utah. I made both of these approximately weeklong trips thanks to Eben Carsey, a Haitian-mission colleague, pediatrician, biker, skier, and reciter of poetry who organized these phenomenal mountain-biking tours for his friends, reserving the huts for us.

The Telluride–Moab trip took place in the autumn of 2000, during the height of the Colorado aspen spectacle. The hillsides were ablaze with golden leaves against the vibrant blue skies at high altitude. It was delightful to cycle up, down, and along the dusty forest service roads, inspired by the beauty around each curve. We were a half dozen bikers, each pedalling at our own speed, but meeting up at viewpoints and for lunch stops. Once we had to pause to watch a dark brown bear ambling across the road in front of us. Another time, we enjoyed seeing a gaggle of wild turkeys gobble their way through the woods, a rare sight in Colorado at that time. Twenty years later, wild turkeys are seen at dozens of sites in the state.

Each evening, we reached a San Juan Hut supplied with fresh water, a well-stocked pantry, sleeping bags, and bunks for six to eight bikers, or in winter, skiers. Our bike panniers contained changes of clothes, a sleeping sheet, bicycle repair kits, and water bottles. How fun to create semi-gourmet meals out of tinned tuna, canned stew, mashed-potato flakes, cheese, mayonnaise, pickles, and other ingredients. Of course, after a day of biking everything tasted great! At night, a vast canopy of stars spread across the heavens, with no light pollution anywhere. Later, we fell asleep peacefully as the constellations moved silently above us.

Each morning we stretched and moaned to get the kinks out of our bodies, revelled in fresh hot coffee and breakfast, cleaned up the cabin, and loaded our bikes with lunch, snacks, and water. We checked our maps and roads for the day, full of enthusiasm and anticipation of the mountains and sights we would see. We rarely saw other bikers until we reached the La Sal Mountains and the Moab area, which is known for its hard-core mountain-biking challenges. We rode into town thirsty for a good cold beer, but since we were in Mormon territory, beer only came with a meal. No problem: we were as hungry as the bear we had seen. We sat down to a great meal and the coveted beer to celebrate our 215 miles of butt-busting biking.

A few years later we tackled the Durango to Moab route. It was shorter, 195 miles, but was described in the guidebook as "difficult." The views were spectacular, but the challenges were sometimes gruelling! I was incapable of either riding my bike or carrying it loaded over boulders on one section of the trail. Fortunately, a young, strong member of our group came to my rescue and maneuvered first his bike and then mine over that stretch! I felt vulnerable riding downhill over rough terrain where I might fly over a cliff at any moment. I opted to push my bike down these goat trails, braking all the way. I knew at those times that this sixty-five-year-old grandma was way out of her league!

We all suffered from sore seats from bumps on the rough trails. But we were inspired by the rugged peaks above us and the vast vistas down long valleys to a pastel horizon beyond. There is a grand feeling of accomplishment when stopping for lunch under the shade of pine trees by the side of the road after a morning of pedalling. We stretched out tired legs, and opened our backpacks for a tasty tuna sandwich washed down with cold lemonade. Then we took a short siesta before hitting the road for an awesome afternoon.

What was meant to be our next-to-last day included a five-thousand-foot vertical climb up switchbacks into the La Sal Mountains of Utah. As we reached the culminating point of our climb, a dark cloud swooped over the crest of a nearby mountain, engulfing us in smoke. The sun turned red. We were just one mountain away from a forest fire. What to do? Should we redescend the five thousand feet we had just climbed? But then where could we go? Our hut for that night waited only a few miles ahead. As we headed for it, we tried to determine which direction the forest fire was moving. We could not see any flames and did not sense any imminent danger. We entered the hut and cooked up a big pot of stew. The smoke did not seem to be getting any worse, but the more we thought about it, the more we knew we could not sleep with a fire in the neighborhood.

It was getting to be dusk as we closed up the cabin and headed for a paved road several miles away that would take us away from the fire. We reached the road as night fell. We had brought with us only two headlamps, as we had not planned to bike at night. One member of our group rode first with a light, and the rest of us followed with a light in the rear. It was a nerve-wracking downhill ride. I was somewhere in the middle of the pack, following the person in front of me, trying to stay reasonably close without tailgating. I couldn't tell how fast we were going in the dark and was nervous beyond knowing.

Eventually we arrived at a roadblock. The State Patrol bawled us out for biking in an area closed for a forest fire. We explained we were biking to escape the fire. They did not fine us and sent us on our way down the mountain road. There was no traffic as we cycled on in the dark. I just prayed we wouldn't crash in our precarious procession. Our road led us to a major highway into Moab. At least we could see by the lights of the passing cars as we hugged the side of the tarmac. At the edge of Moab, we found the motel Eben had reserved for us for the following night. We explained our predicament to the owner, who had several extra rooms available.

It was too late for dinner in a restaurant. We pooled our extra resources and shared PowerBars, apples, and hot tea for a late supper. We were still trembling from the emotion of our night ride in the dark, but fatigue won over nerves as we eventually fell asleep. So ended my very last mountain-biking adventure. I donated my mountain bike to a foreign student at the university who was looking for transportation. He passed it on to another student when he graduated. Who knows where the bike is now, as I both shudder at and enjoy memories of the La Sal Trail, my last mountain-bike ride.

Figure 36. My bike loaded for the trip

Figure 37. Mountain biking on a forest road

Walking around the World

According to my Fitbit, I have walked the distance of the diameter of the earth over the past several decades: eight thousand miles. Over my lifetime I have certainly walked considerably more, just guessing, but at least the circumference of the earth. From an early age, walking was what I did. The summer I was six, my mother sent me on foot across town three days a week to the Hygienic Swimming Pool, purified with gallons of chlorine. I walked two miles to the pool to learn major strokes and swim multiple laps before I trudged back home another two miles.

Of course, my siblings and I walked to school and back during all our time in public education. Though I never endured life-threatening snowstorms or had to slog dozens of miles through blinding blizzards, I still carried what felt like tons of textbooks to and from the classroom and the homework table. Walking was a way of life in Boulder, Colorado.

Since then, I have continued to be an avid walker, not only for exercise, but for everything I see and experience in the process. In my college days, I spent one summer in New York City walking to work through Chinatown, an Orthodox Jewish district, and warehouse areas where for the first time I passed homeless men sleeping on the streets. Later, and on another continent, I inadvertently strolled through a small park in Frankfurt, Germany, where drug addicts were heating and injecting heroin. I felt as if I had dropped into hell on earth.

In Italy I hiked for two days with special friends. We walked suspended on trails across cliff faces on the Via Ferrata, the Iron Way, retracing the movement of Italian troops in the Dolomites during World War I: spectacular and historically intense! In Norway I trekked with Elisabeth, our Norwegian exchange student, on tundra trails across vast highlands to overnight huts. There I met other hardy Norwegians in love with exploring their county. I loved doing it, too! In Sweden, exchange student Freddie walked with us through rocky areas covered with petroglyphs of Viking boats and symbols left by her forebearers. Walking in rural Turkey, I looked down and found a Roman coin in the dust.

When the Berlin Wall came down, Hajo and I met a couple from Dresden. For the next ten years we walked with our friends Hans-Jorge and Ulli across worlds they had never been allowed to visit. We cemented a deep friendship as we strolled together through Paris and London; hiked though parts of former West Germany, Austria, Portugal, Alsace, and England; and explored US national parks and Hawaii on foot.

I shared my love of walking with my grandchildren, Brendan and Kayleigh, when they were quite young. We made multiple backpacking trips on the Olympic Peninsula, hiking on boardwalks to sandy beaches on the Pacific coast. These annual jaunts have been repeated years later, as Kayleigh now leads me on annual camping and hiking adventures to old and new spots.

Nepal

Of all the walking and hiking experiences I have enjoyed in my life, one stands out as uniquely special: twenty-three days of trekking the Nepalese Himalayas in 2001 with Mark. It was the ultimate walking adventure, from the village of Lukla at 9,500 feet elevation to Everest Base Camp at 17,500 feet, with side trips to extraordinary villages and valleys. Weeks without television, telephone, internet, or world news. Weeks of sharing trails and memorable times with local villagers, Sherpas, porters, international climbers, and group members. Our trek was organized by Phursumba Sherpa, our guide on Mount Rainier, who invited us to join a group of hikers to explore Nepal with him. It was not a walk in the park, as several ladies discovered early on. They left our group to join a less strenuous program.

We flew into Kathmandu and were greeted at the airport by Gombu Sherpa, who had led Jim Whitaker, the first American to summit Mount Everest. He draped white welcoming shawls over our shoulders and escorted us to our hotel in the center of Kathmandu. En route, we stopped to visit the famous Monkey Temple, where furry monkeys frolicked on the walls and trees around the complex. This was March 31.

I remember the date because the next day was April 1. Our family was always notorious for practical jokes on this day. First thing in the morning, I confronted Mark in the hotel hallway with a problem. "Mark, there's a monkey loose in our bathroom." Mark rubbed the sleep out of his eyes as I explained, "The monkey must have found his way over from the temple. We left our bathroom window open. How do I get rid of him?" Mark looked exasperated, as he didn't know what to do. I resolved the issue with a "Happy April Fools' Day!"

Our walking trip began at Lukla, a primary trailhead for Mount Everest. Our group of a dozen hikers and as many porters and Sherpas joined throngs of people navigating the trail, including locals walking from one village to the next or going to work in potato fields carved out of the rocky landscape. Porters walked with incredible loads, which were often balanced precariously and held by wide cloth bands over their foreheads. I was most impressed by the egg carriers: men with hundreds of fresh eggs loaded in wire mesh containers strapped to their backs. One stumble and a massive omelet would have spread over the trail, never to reach hungry hikers in the guest hostels for whom the eggs were destined.

Our group preferred staying in tents pitched by our porters in spectacular spots along the trail. No motor vehicles could navigate our meandering dusty paths or the hundreds of rocky stairs, or cross the deep gullies on decrepit, swinging suspension bridges lined with prayer flags as insurance. Even with the flags, one yak bearing a heavy load lost its footing between the bridge planks and the trail, dropping to its untimely death.

Everything in the Nepalese highlands is transported by men or beasts, on foot or hooves. Our trekking group had a woman dzos driver who was in charge of six dzos carrying tents and equipment. Dzos are a cross between a male yak and a cow. They are considered easier to control than yaks. Our dzos driver had to load and unload her charges each day and drive them to the tent sites. One afternoon, a contrary dzos that was just unloaded took off down the trail for home. The driver had to literally run after it. One day later she caught it. Grabbing a branch, she switched it all the way back up to our camp, where it continued to carry our gear for the rest of our trek. Just one example of how difficult life can be in Nepal.

Yaks and dzos not only carried goods. Their dung heated homes and cooked meals for many settlements above timberline. As we strolled through villages, we saw patties of yak dung plastered on walls to dry in the sun for later use as fuel. The vile odor of dung smoke rising from village chimneys did little to enhance the atmosphere of our trek! In addition, the dust of the trails was sometimes so impregnated with dung dust that everyone wore bandannas across their faces like bandits to keep from breathing the unhealthy mixture. We all gave plenty of leeway to the loaded yaks and dzos, which moved up the trail beside us with their sharp horns and clanking bells. It was these bells that often woke us in the early morning as small caravans of yaks passed near our tents. Yaks were part of daily life for both the Nepalese and us trekkers.

Walking the trail toward Everest, I was constantly reminded that Nepal has a deeply spiritual background. Omnipresent prayer flags dotted the landscape, waving from trees, bridges, and whitewashed shrines called chortens with a central spire of gold rising to the heavens. In the center of the trail, we occasionally encountered mani stones—clusters of rocks with engraved flat surfaces depicting symbols and deities. We were directed to pass around them on the left to assure good luck.

We visited several monasteries of varying sizes and significance. In Khunde, the monastery curiously boasted of having the skull of a genuine yeti, the Nepalese equivalent of the American Sasquatch. We examined a hairy head locked in a glass case and wondered. In a monastery above Thame, hundreds of little boxes were encased in a wall, each with an ancient scripture inside that was studied by present-day monks. On a nearby river shore, brilliantly painted deities on a low cliff seemed to bless a raging torrent.

At Tengbouche, I visited a renowned Buddhist monastery. The site was awe inspiring, with soaring icy peaks rising on all sides above the village. The monastery had to be reached on foot after days on the trail. I entered the austere, maroon-colored structure and walked up the cold stairs to a dimly lit room. In the center, an altar glowed, with clusters of burning butter lamps illuminating Buddhist statues and chanting monks in dark-maroon robes. Nepalese men and women in coarsely knit woolen garments, Western climbers in the latest brand-name gear, porters in worn hand-me-downs, and a few tourists and other folks all sat or knelt on concrete platforms around the altar. I joined them in the back, soaking in the spirit of the setting.

The monks who had been chanting soon fell still. We all remained in a meditative silence for many minutes until suddenly, I was startled by the crashing noise of cymbals, drums, and cries. Then all resettled into a sublime stillness. Bang! Another burst of ghastly noise. Then came more quiet time, which changed to chanting as the monks slowly filed out from around the altar into the darkening sanctuary, leaving the still-flickering butter lamps alone. Only later did I learn that the crashing noise was to call the people in meditation back to awareness when they started to drift off in the silence. I was so inspired by the service's dramatic setting, the chanting, the glowing flames, and the sense of ages of deep spiritual tradition still alive, that I was very much awake and filled with gratitude for this awesome interlude amidst our challenging trek.

Mark and I cherish so many highlights of our days walking through Nepal. Children in many layers of colorful clothes scurried past us on their way to school. We visited several of their minuscule classrooms, where they crowded together on rough wooden benches to learn math, reading, and songs in English from smiling Nepalese teachers. Many of the schools were supported by Everest climbers. In Khunde, our trail passed by a clinic where a New Zealand doctor and several nurses tried to meet the health needs of ten thousand villagers in the region.

Even though we were trekking in April, as we reached elevations above fourteen thousand feet, it often snowed at night. We awoke one morning to bright sunshine and six inches of fresh snow. The porters were busy constructing a dozen obviously naked snow "men" around the campsite. They were laughing and joking with delight at their snowy creations.

Everywhere, we were walking in the presence of mighty mountains. Some were stark and terrifying in their sheer size. Ama Dablam was like a beautiful woman, her curving outline covered in white gauze. But it was Mount Everest that commanded our attention.

Mount Everest

Mount Everest, at 29,035 feet, is celebrated as the highest mountain on earth. It is honored in Nepal as Sagarmatha, Peak of Heaven, and in Tibet as Chomolungma, Goddess Mother of the World. Its first documented climb was made in 1953 when New Zealander Edmund Hillary and Sherpa guide Tenzing Norgay made their record climb together, assisted by a full expedition. I was in high school at the time, making my first semi-technical climbs in the Rocky Mountains. I secretly imagined someday summiting Everest myself.

Forty-eight years later, Mark and I hiked together into Everest Base Camp at 17,500 feet. We were relatively acclimated after walking our way up the Khumbu Valley. We were excited to be in this staging area for so many climbs of Everest, still twelve thousand feet above us. As we looked up at the Khumbu Icefall directly in front of us, we lost any interest we may have harbored in climbing Everest. We could see parts of a tortuous route of ropes and ladders zigzagging up the ever-moving glacial ice cliffs, blocks, and crevasses. One could never know when they might shift or avalanche, taking out climbers or Sherpas. Above the icefall lay the challenges of more avalanches, crevasses, violent storms, altitude-induced illnesses, and the long lines of climbers hoping to reach the summit during the short period of favorable weather. As we explored base camp under a canopy of prayer flags, we remembered that hundreds of Sherpas and climbers had lost their lives on this mountain.

Considering all these facts, there was still a palpable sense of energy and optimism in base camp. We talked with some climbers making their first summit attempts and others their third and fourth. Some had completed multiple successful climbs and were back for more. The Sherpas were among this last group. Their jobs depended on it. I remember standing in awe beside Babu Chiri Sherpa, knowing he had climbed to the top of Everest in seventeen hours from base camp, had spent a night on the summit without oxygen, and had guided multiple clients to the summit over the years. He hoped to retire eventually and organize a school for village children. A few days after our conversation, he fell into a crevasse and died.

We left base camp with a mixture of emotions. We were exhilarated at having reached our destination and having shared the enthusiasm of the climbing season just beginning, awed at the grandeur surrounding us, and sobered by the ever-present odds of accidents and death in attempting to conquer mountains that a person never truly could. They could summit them, perhaps, but that was all.

Our final days in Nepal were spent on long, pleasant walks down trails we had climbed earlier, but now we walked through spring flowers and rhododendron. We made interesting forays into valleys that looked like Shangri-La. We had hoped to climb 20,300-foot Island Peak, but our attempts were thwarted by a storm a few hundred feet from the summit. Overall, the Nepal trek was a grand walking adventure with Mark, about which we reminisce often.

Walking and Volunteering in Colorado

In between overseas excursions, Hajo and I made Boulder our home. We both volunteered for a trail-building venture in Arizona on the Continental Divide Trail that runs north from Mexico to Canada along the Continental Divide! It was one of the few stretches that was poorly marked over volcanic flows. As we worked, I began thinking about making one of these long walks myself. The Appalachian Trail was too far away. The Pacific Crest Trail too physically demanding. The Continental Divide Trail simply too long, at over three thousand miles.

Then I heard about the Colorado Trail, just under five hundred miles. That seemed doable. When Hajo and I returned to Boulder, I talked with some of my hiking-group friends. They said that they worked on a section of the Colorado Trail every year, keeping it in good shape. I signed up to help them and fell in love with the trail.

The Colorado Trail runs across the state from Denver in the east to Durango in the west, crossing mountain ranges, dipping into forest valleys, weaving over flowering tundra, passing by pristine lakes, and presenting breathtaking views from one mile to the next! I had walked on parts of it doing trail work. The most memorable volunteer week was repairing a part of the trail for the Leadville 100, one of the most demanding ultramarathon races in the world. Runners complete one hundred miles at high altitude, starting from Leadville, Colorado, going over Hope Pass and down the other side to the bottom, then ascending back over the pass again and finishing in Leadville.

At the top of the pass, we found a badly eroded chute, which our group was to repair in the week preceding the race. We were a dozen volunteers and four paid forest service workers. Our job was to close the chute, filling it with rocks, soil, and plants we would take while cutting a new switchback section of trail. This would stop erosion in the gulley and give the runners safe footing for descending on a firm new trail.

We volunteers backpacked up to a meadow a few hundred vertical feet below the top of the pass. We were carrying our food for the week, along with tents and sleeping bags. I fell in step with a Japanese woman carrying a pack half as big as she was. A heavy cast-iron teapot was attached to the back. She told us she was a forest ranger in Japan and wanted to see how volunteers worked in the US.

She and I pitched tents next to each other. I set about fixing supper, and asked what she had brought and if she needed anything. She said she had brought seven rice balls, one for each day. She had gained so much weight in her two-month stay in California that she had to lose weight, and she had decided this would be the best way. She heated up some water for tea and ate her rice ball. I told her I had brought plenty of food and would share it with her. She would need to have more than a rice ball's worth of energy each day for the work before us.

The next morning, we were up with the sun, eager to begin our trail work. One of the rangers gathered us volunteers and explained that the chief forester was "indisposed" and unable to join us. But it was clear what we had to do. We each carried a trail tool and our pack with work gloves, water, snacks, and rain gear to the top of the pass. For two days, we cut tundra for the switchback and carried what we had removed to fill in the ravine at the top. It was good hard labor, and it was fun to work together.

On the third day, I stopped to take a breather and looked down the trail. A solitary figure was steadily making his way up toward the pass. He was dressed all in white, with a white sombrero, white gloves, white pants and shirt, and white high-top tennis shoes, with white sunscreen dabbed on his face. My first reaction was one of disbelief. What was this!? Then I said to myself, "Do not judge. Do not judge." The paid forest service people went over and greeted him warmly as he approached us. He was the head of the project. He called for a break and congratulated us on the job we were doing. Then he said, "You are probably wondering why I look so weird?" We hesitated to laugh as he continued, "This is a story of doing everything wrong!"

He explained he had decided, during a break from his Hope Pass work, to spend a few days hiking in the Hells Canyon area on the Idaho side. He didn't tell anyone where he was going. It was a spur-of-the-moment decision. He took off in his car from Colorado, drove to Idaho, and camped out somewhere near a trailhead. The next morning, he threw stuff in a pack, picked a trail, and took off. He was wearing jogging shoes and having a great time as he headed down toward the Snake River. He had been hiking most of the day when he came to an area of thick ferns.

Suddenly, he felt a stabbing pain in his ankle. He looked down to see a thick rattlesnake coiled, hidden in the ferns. He knew he'd been bitten and he was in trouble. Why hadn't he worn high-top boots? He made a tourniquet, wrapped it around his leg, and tried to force the venom out of his wound. He was starting to feel really ill, still miles from help. He decided to go to the river. It couldn't be too far. Maybe he could hail a boat going upstream.

He made it to the river and found a small group of river floaters. They really wanted to help him, but they were drifting downstream and were days away from help, going south. They could not go upstream without a motor. They tried to make him comfortable for the night. The next morning, he felt somewhat better. He decided his only choice was to walk out the way he had come in.

It was slow going. The further he went, the sicker he felt, and the weaker he was becoming. He knew the venom was all through him. He started caching things out of his pack: his tent, his sleeping bag, and made walking sticks to hold himself up as he staggered on. He left his pack hidden behind a rock. He was afraid the end was coming. Finally, he was literally crawling along the trail.

Then he heard voices. A couple was coming toward him. At first, he thought he might be hallucinating, but they were for real. They said it was about three more miles to the trailhead and his car. He knew he couldn't make it. They said they had a radio and would try to call for help. He felt his first glimmer of hope. The call worked. Within half an hour, he was being lifted into a helicopter. Within an hour, he was in intensive care in the hospital. Just in time! His organs were all shutting down. It was too late to use antivenom. He didn't know everything they did to save him, but they did. His parents came to the hospital. They were grateful to see him alive, but gave him a good scolding for everything he had done wrong. He felt the whole episode was a miracle and a wake-up call.

He spent a few days recovering before the doctors said it was okay for him to go back to work on Hope Pass, with certain limitations. He was taking medications that required that no sun touch his body, hence the sombrero. He could work as hard as he felt like, but he wasn't to force himself to do more. His story complete, he told us the break was over. We all went back to work.

During the following days, the sombrero was easy to spot. Our chief worked with us, then offered us free-time excursions and activities. One afternoon, he led us up to a high ridge above timberline, where a lone miner's cabin clung tenaciously to the tundra with shallow mining pits dug all around it. There was no source of water, no wood for fire, no neighbors. Why was it there? How could anyone have lived under those conditions? Another day, he led a hike up Hope Peak at almost fourteen thousand feet, because the Japanese ranger wanted to climb a real mountain. The other hikers were much faster than she, but our chief paced himself according to her needs. He stayed with her until they reached the summit, and then hiked back down to camp. He won my admiration for thinking of her rather than taking off with the pack.

I saw the way he hiked with skill and ease and asked him if he had climbed much. He mentioned he had climbed Denali, known as Mount McKinley in those days. That night, we all laid out our sleeping bags in the meadow and lay in them watching a meteor shower. Streaks of light shot across the night sky. Incidentally, I shared my supper with the Japanese ranger—her rice balls were long gone!

We finished our trail work right on time. The switchback was complete and smooth, the gulley replanted like a rock garden and roped to show that it was off-limits. We were packing up to leave when the first of the Leadville 100 runners passed our camp area. They had already run more than a marathon from Leadville and seemed fresh and relaxed as they took a short water break. I could not believe their conditioning!

A few years later, I was doing my walk of the Colorado Trail. The Hope Pass part I had worked on was an optional segment. I had decided not to do the walk alone, but with other hikers organized by Colorado Mountain Expeditions. They moved our tents, sleeping bags, and gear for us and fed us fantastic meals. They divided the CT into seven segments of one week each. They suggested we do no more than two segments each summer. It took me five summers to complete, with everything else going on in my life. The memories of these treks fill several photo albums. One of my favorite trail companions was a tenacious teacher from Jamaica. She came from sea level every summer to a ten-thousand-foot-high location to hike the Colorado Trail. We crossed the final line together!

Remembering Verlie

I have encountered many remarkable people in my travels, but Verlie, whom I met in a care facility near Boulder, stands out. Every week for many months when I was not travelling, I would drive some fifteen miles to bring her Communion from our church. This was before Covid, but even then, Verlie was unable to go to church. She was bedridden with the recurrence of polio symptoms.

As a child, she had contracted polio and spent part of her youth in an iron lung, and she had spent most of her life on crutches. Now she lived her days propped up in bed and looking out one window. She shared the room with another woman behind a curtain who slept most of the time. Despite the circumstances, Verlie was always positive. She expressed her perspective in the following poem she wrote and shared with me.

A View of Space
Some think I am confined
In this half of a room.
On some faces I read
They think it's a tomb.
Don't they know the magnificence,
The power of thought?
They ought!

If this was all I had to remember her by, it would be enough. We developed an unlikely friendship. I was traipsing around the world while she was barely moving in a hospital bed. Yet, we both cared about each other and the world. She always wanted to hear about my travels; showing a genuine interest and asking pertinent questions. She was a care receiver, but also a caregiver. Verlie personally knew those who helped her every day, and would ask me to get special gifts for the children of her caregivers to show her appreciation.

I never heard her complain about anything. One particularly loud patient would cry out for whatever reason. She would say, "I'm so sorry that Mr. Jones is having a hard day today." A director of the facility would consult with her about various issues in the institution because he respected her and her insight and experience. She loved to play Scrabble. As hard as I tried, I could rarely beat her, and we both enjoyed the competition. I hope that if I become as dependent on others as she was that I may be as optimistic, kind, and caring as she.

Very gradually, I learned about her life, which made me wonder even more about her optimism. She grew up pretty much a ward of the state, living in the hospital with polio as a child. Even then, she decided she wanted to be an educator and work with children in the hospital, like her own hospital teacher. She finally became well enough to attend university on her own with crutches, and she became a student teacher of English in a regular high school.

At the end of her student teaching, she asked her principal to write a letter of recommendation so she could become a hospital teacher. Instead, he wanted to hire her to teach in his school. He said that she was a role model. She demonstrated that despite adversity, a person could succeed. Verlie said that she enjoyed her teaching and loved her students. They soon saw that despite her physical limitations, she meant business and wouldn't let them get away with nonsense.

During her teaching years, Verlie married and had a son. She didn't say much about her husband and son, but they were deceased. However, her son had raised a daughter, her granddaughter, Jennie, who was a student at Colorado University. She was the one who had helped Verlie find the care facility nearby. Jennie was the delight of Verlie's life. She was the one who contacted me when I returned from teaching in Japan to tell me Verlie had passed away. I missed her. I missed her funeral and her celebration of life. Jennie left me two of her poems. They express the spirit of this remarkable woman.

BINGO

They come to play like children;
To add to the din;
To possibly win;
To eat; to talk; to laugh;
To bravely conceal; or
maybe reveal
the hearts of the heart
the flesh;
the bone;
Because it is better
Than being alone.

THE TECHNIQUE

My friend moved away;
So when this thing happened
I went to the psychiatrist.
He was very nice.
He listened for a whole hour
And never said a word.
Just like my friend,
Only, she served cookies and tea
And never charged me anything.

From Southern Hemisphere to Svalbard and Back

Mishaps Make the Machu Picchu Adventure

Machu Picchu had been on my bucket list for many years when I discovered a small local tour company near Boulder that organized tours to this Inca wonder in Peru. It was 2006 and Hajo and I had returned from one of our teaching engagements in Europe, and I had some free time for a tour while he was involved with other responsibilities. I contacted the company and said I wanted to hike the Inca Trail, explore the site, and visit others in the vicinity. The tour operator said they had several other clients wanting the same kind of tour and set a specific date and itinerary that met my expectations.

As the date approached, I asked for verification of the details. A day later, they contacted me. Our reservations to hike the Inca Trail had been scheduled for one month later than our actual tour. Whoever had made the mistake was unknown. There were no more spots left to hike the traditional Inca Trail. But, if I didn't mind, we could hike a lesser-known but more spectacular Inca Trail, one of many that led to Machu Picchu. I decided to go for it.

Two days before our intended departure, the tour operator called to say that the other tour participants had cancelled. Would I be willing to be the only person on a kind of reconnaissance tour? We would follow the same itinerary but explore using llamas, horses, and human porters to see which would work best for upcoming tours. I agreed, but wondered what kind of tour this would be. The local American guide, whom I will refer to as the "leader," said he would meet me at the airport in Cusco to begin our tour. I used accumulated mileage to obtain tickets. I flew into Cusco right on time and

Mishaps Make the Machu Picchu Adventure | 247

met a tour representative, who took me on a local bus into this city at eleven thousand feet. We spent several days there acclimating and visiting impressive local sites.

Then came the day to begin our Inca Trail trek. I clambered into a small van with our American tour leader, a Peruvian guide, two group porters, and a cook with a huge bag of cooking gear, tables, and tents. We drove to a nondescript trailhead and met a llama handler and two llamas, a kind of cowboy, and two horses: an expedition of seven support personnel and four animals for one tourist—me! Above us, in the distance, the spectacular Andes rose with impressive glaciers and summits.

Our goal for the day was to reach a valley that cut up and through this cordillera. We would stop halfway along for a lunch break. The handlers loaded the horses and llamas with part of the gear. The porters took the rest. I was left with my day pack. We all trekked along at different paces, and were soon strung out over a mile. I stuck with the Peruvian guide, who spoke limited English (but it was much better than my Spanish) while the American tour leader tried to coordinate the whole procession.

When we stopped for lunch, the cook prepared a quick meal while we relaxed. We ate before hitting the trail again in various stages. When I reached the designated campsite for the night, the tents were already pitched. Everything else was in a huge pile on a tarp. We had started sorting when our American leader declared he had left his new parka on a tree back at the lunch stop. He would have to hurry back and get it. We were left to our own devices as the cook looked for the stove, pans, food, and table. At the bottom of the heap, I found my duffle with my sleeping bag and all my accompanying possessions. I set them in my tent and zipped it shut.

I slipped away from camp up a trail to take in the evening sun on the mountaintops. To be alone in all this grandeur was truly overwhelming! By dusk I was back in camp. Our leader had not yet returned, but the cook had a big pot of soup. We all sat around slurping. The porters, animal handlers, and the rest of us were one happy group. It was dark when our leader returned without his coat. He said that someone must have found it and needed it more than him! He would manage.

The next morning, we set off hiking up the trail to a high mountain pass at fifteen thousand feet. The air was thin and cold, the trail challengingly rough, the views spectacular. We all stopped on the top of the pass, except for the horses, which had gone ahead. I tried to make friends with the llamas, but they spit at me, which I had heard could happen. Still, they posed with me for a photo on the top of the pass. The rest of the Andes rose thousands of feet about us. The usual Inca Trail could not have been this breathtaking!

We crossed the pass and descended into a steep tundra area. I was hiking with the local guide when he told me he had grown up in this very valley. He still had family living in a little village we could see way down at the bottom of the incline. The trail cut in switchbacks when we came to a spot where the guide said, "Stop! I think I see my cousin over there working on the hillside. Let's go see him!" We left the trail and headed across the steep slope. It was his cousin, and family, and a couple of other men. They were digging in the tundra at about fourteen thousand feet to plant potatoes. They thought that with the climate warming, they would be able to harvest a crop at this altitude. Maybe global warming has a good side?

The next days were a jumble of impressions. We stopped at one spot for a dip in a small hot spring beside the trail, a luxurious, cleansing, relaxing break. As we lost altitude and it became warmer, the llamas stopped cooperating and the handler and llamas returned over the pass, their loads redistributed to the horses and porters. The vegetation became junglelike, and we came to a small town with stores, everything transported there by men and beasts. At this point, the cowboy and horses left us for some reason, taking some tents and other things we wouldn't need. The trail continued widening out the further along we hiked, until it intersected a

Mishaps Make the Machu Picchu Adventure | 249

road where trucks had stopped and were being unloaded. They had brought materials to build a hotel for tourists wanting to trek this Inca Trail.

We left the road and returned to the trail, where we began to climb again. By the end of the day, we reached the beginnings of another tourist hotel for trekkers. Here, our two porters decided to leave us. They could make more steady income working to build this hotel. We were down to four adults and had more than we could carry.

Two local children approached us. They wanted to be our porters. They were about twelve years old, a boy and his sister. Our leader agreed to try them. I was upset. Wealthy American woman using child labor in the third world! However, it turned out to be great fun for everyone. We adults carried what the children could not carry comfortably. They laughed and joked on the way up the trails. We stopped for popcorn breaks. At night, they slept in our parkas and other clothes in makeshift sleeping bags. They helped us and we helped them.

Nearing the end of our trek, we reached a high ridge with an immense area of ruins half-covered with vegetation and roots. Looking across a very wide, steep valley, we could see Machu Picchu far in the distance. Our local guide said that the ruins where we were to camp that night were of a city built at the same time as Machu Picchu but was not yet excavated. We could explore, but were not to touch or remove anything. Our porter children were excited to have this opportunity! It was amazing to be able to be in the midst of such a remarkable ruin, much like what Hiram Bingham had rediscovered at Machu Picchu in 1911! This Inca Trail was indeed remarkable!

The next morning, we all scrambled down the steep trail to the valley bottom. I remember the cook's pots and pans clanging as he maneuvered over the deep steps, while the children bounded carefree with their loaded packs bouncing on their backs. At the bottom of the trail, we came upon the tracks of the train line that ran to Aguas Calientes, the tourist town at the base of the road and trail up to Machu Picchu.

Here we split ways. The local guide and cook would take the children to nearby relatives, while the leader and I would take the train to Aguas Calientes. I made sure the children received good tips!

The tour itinerary included three full days at Machu Picchu. We stayed at a hotel in town and either hiked or took a local bus up the steep hill to the restored city. The first day, the leader gave me an orientation tour of Machu Picchu in its dramatic particulars. The next two days, I was on my own to explore at my leisure with a detailed guidebook. It was wonderful to imagine the ceremonies and way of life at the height of the Inca empire. I was awed by the setting, the original stone workmanship, and the authentic reconstruction. These impressive details are too numerous to outline here.

I took time to climb to the top of Huayna Picchu, the steep mountain directly above Machu Picchu. Until I was there, I did not realize that this whole mountain was covered with structures clinging to the cliffs. Hundreds of steps led me up and around the mountain. At other spots, I had to tackle steep dirt trails with lines of other tourists. We climbed gingerly, just hoping a domino effect of falling bodies would not take us out! This was not only a sacred mountain, but also an unrivaled viewpoint overlooking the whole Machu Picchu complex!

I loved my day spent hiking down and then back up the last part of the original Inca Trail. It was amazing to reach the Gate of the Sun and catch a first view of Machu Picchu! On the nearby hillsides, I discovered numerous plants and flowers I had never seen before. I appreciated the many, many steps that comprised the Inca Trail and was happy I had taken our alternative trail! My American leader took me to the other major tourist sites of this part of Peru, and we ended our tour attending the wedding of one of his Peruvian friends. This was indeed a tour unlike any other, and just right for me.

Pat in Patagonia

Patagonia is the area in the far south of Chile and Argentina. I thought of it being almost in Antarctica, but it is really at about fifty degrees south, about the same latitude south as the US–Canadian border is north. Strangely enough, the border between Argentina and Chile at that latitude resembles the border between North Central Montana and Canada: flat dry prairies, gravel roads, incessant wind, and a swinging metal gate with a padlock. I hadn't travelled thousands of miles to see what I had already experienced for twenty years in Havre. But this was the only déjà vu experience of my three-week adventure in March of 2008, early spring in the northern hemisphere and late summer in Patagonia.

I love travelling south from the US rather than east or west because there is little to no jet lag! Though seeing the noon sun in the north takes some getting used to. The peaks of Patagonia are spectacular beyond imagining, and it was these mountains that drew me there, like the Matterhorn had drawn me to Switzerland fifty years earlier.

Patagonia in Chile

I flew from Colorado, and like with my trip to Peru, I was informed at the last minute that the group I was planning to trek with had backed out of their tour. So, again, I was a tour group of one with a fantastic guide, Eduardo, who shepherded me through his homeland.

The first day I spent exploring Puerto Natales, Eduardo's hometown, while he searched for and found my luggage, which had been lost en route from Santiago. He told me where to look for the local black-necked, red-beaked swans with white bodies that were typical of the area. Really eye-catching out in the bay! I talked with boatbuilders who were making small craft to ply the waters around Port Natales. I visited a butcher shop and found that meat is an essential part of meals in Patagonia. I discovered an orphaned young guanaco, a wild form of llama, who was being cared for until she could return to the wild. Her long eyelashes made her look flirtatious.

The next day, Eduardo and I headed off to the Andes and the start of our trek. Along the way, we passed herds of guanaco and groups of nandu, ostrich like birds, running across the prairie. I was amazed to learn that the male nandu builds a nest for the female. She lays the eggs and leaves them for the male to incubate by sitting on the nest for about a month until they hatch. Then it is he who takes care of the young. Quite a separation of duties in which the female is the winner!

As we drove toward the mountains, the rugged spires of the Torres del Paine rose ever closer and higher. We crossed a rather primitive wooden one-way bridge over a raging river. Eduardo was concerned that the flooding river might take out the bridge, with global warming melting the glaciers so much faster than in the past. The mountains here are unique in the world because of their formation. They are composed of a white granite laccolith, on top of which lies a swath of pure black metamorphic rock. These two formations have been carved by glaciers into dramatic spires and towers over the eons. Glaciers are still at work in some places and have entirely cut away the black summits, leaving white cliffs and monoliths rising above numerous lakes.

We trekked around several massive ranges for most of the week. We were fortunate because the weather was sunny and warm most of the time—unusual for an area known for wind and rain and snow on the summits. We camped by Grey Lake, where global warming had melted a major glacier back by a mile in the last fifty years. I talked with two ecologists who were photographing the glacier, the icebergs, and the expanding lake to try to show the world the realities of global warming.

Later, Eduardo and I met a group of trekkers preparing for an afternoon happy hour. They said we could join them if we brought the ice for the pisco sours. Eduardo and I went down to the lake and found some small blocks of glacial ice, which we chipped for our drinks. That ice might have been one thousand years old. No way to tell, but it was unforgettable in our cocktails!

It is hard to put into words the incredible wonder of seeing such a variety of splendid mountains, one after the other. By walking the trails, a trekker takes in the total experience of this wild, pristine world through the rhythm of walking, the feel of the trail underfoot, the inhaling of clear cool air, the viewing of condors soaring and circling above, the tasting of pure water taken right from the streams, and the warmth of the sun on shoulders. One evening before dinner, I wandered out to a ridge where I sat in silence, mesmerized by the play of sun and shadows on lofty summits. I jotted in my notebook: "Sunny, dark, misty, stark, Creator and creation in a concert of oscillating clouds, cheered on by an audience of one!"

The next day, we took a catamaran across Lago Pehoé and saw our last views of the Cuernos del Paine. This introduction to Patagonia was more rewarding than I had ever imagined. We were only halfway through. The Patagonia of Argentina waited ahead of us.

Patagonia in Argentina

Many of the local inhabitants I met in Patagonia seemed more like Patagonians than either Chileans or Argentinians! They flew the Patagonian flag right beside the national flag. Eduardo, as the leader for the whole tour, accompanied me to Argentina, where I was required by law to have an Argentinian guide as well as my local one. Two guides for one client, plus a porter to carry my pack. I wasn't too spoiled! The Argentinian trek centered around the Fitz Roy Massif, which I had never heard of until I arrived on the scene. I was overwhelmed by the massive towers and pinnacles rising above vast glacial ice fields that spread across and around the Southern Andes.

My local guide was Mario from Mendoza. He carried a blue guitar attached to his backpack. He said he had a reputation to uphold. The men of Mendoza were known for serenading the ladies wherever they went. Several evenings after dinner, he serenaded me as we relaxed after a day on the trail. I was charmed! Mario was also a technical mountain climber. He looked at the various summits, contemplating possible routes he might someday scale. He told wrenching stories of climbers dying from ice falls and horrendous storms. I remember him with trepidation, wondering if he is still surviving to serenade the ladies.

He knew other tour guides in the area and suggested we join another group for a day of ice climbing, if I was interested. Why not! First, we had to get on the ice field, which was across a turbulent stream. Someone had anchored a metal cable across the river. I watched Mario cross it as if on a playground contraption. I had to clip myself to the cable with a carabiner attached to a harness around my waist. Wrapping my legs around the cable and leaning back, I let the carabiner hold my weight. I pulled myself across the cable hand over hand. Ingenious! I made it.

The ice field was a magic world. Gleaming white ice with dark crevasses of deep blue stretched up to the base of vertical granite walls. The undulating ice fields were dotted with meltwater ponds, some blue and some olive green. Gigantic blocks of ice had been tipped on end as the glaciers moved. It was one of those steep blocks, seventy-five feet high, that we would attempt to scale. The group we joined had already climbed when we arrived. A belay rope was ready, so we hurriedly attached crampons to our boots.

Mario handed me a special ice ax for each hand. He demonstrated how to engage crampon spikes and ice ax points into the ice and work one's way to the top. He moved to the summit with the skill and grace of a dancer, alternating legs and arms in an easygoing rhythm. Then it was my turn. I attached a belay rope in case I fell. I attacked the ice tentatively, one limb at a time, and then got the feel of it. It was really fun jamming spikes into the ice and moving step by step upward. As I reached the top, it occurred to me that you really can teach an old dog new tricks!

After the ice-climbing excursion, we crossed the river again on the cable and hiked back to our base camp. Our tents were pitched beside an area of small, flat, round stones resembling coins—thousands of them! I felt like King Midas, waiting for them to turn to gold! Meanwhile, my guides relaxed, sharing a large gourd of yerba maté, the traditional Argentinian drink infused with caffeine. It has since become popular in the US, and available in many grocery stores.

That night, I was awakened in my tent by the sound of mice chewing loudly in what I thought was a neighboring tent. I groaned to myself, "Who was the idiot who left his tent door open and kept food in the tent besides?" I tried to sleep, but the chewing sound was so loud it kept me awake.

Finally, I flipped on my flashlight, which reflected off two pairs of little beady eyes. They were mice all right, and in my tent, chewing on a bar of soap!

How to get rid of them? I grabbed all my belongings and pitched them out of the tent, then turned it upside down. The mice scurried away in the dark. I stuffed everything back into the upright tent and crawled into my sleeping bag. The last thing I heard as I fell asleep was the raucous call of a screech owl. Perhaps one mouse down?

The Fitz Roy Massif is composed of a series of four sheer granite spires rising out of the ice fields around the major peak, Fitz Roy, itself. Each bore the name of one of the first pilots to fly airmail across the Andes. Among them was Saint-Exupéry, who had written *The Little Prince*, a book I had taught my French language students to read in Havre. I had been on the rocky coast near Marseille when divers found the plane in which Saint-Exupéry had lost his life as an air force pilot during World War II. I had come full circle tracing his life. This Patagonian pinnacle was a fitting memorial for him.

Our trekking trails skirted these peaks, around turquoise lakes, over open tundra, and through lenge forests. Lenge were almost like natural bonsai trees, stunted in growth because of the very shallow soil and harsh weather. Their miniature leaves were turning gold, marking the end of the trekking season.

For several days, we hiked the jagged edge between summer and autumn. I wrote in my journal: "The wind brushes of autumn are painting swaths of gold across the green palette of summer." Mostly, the winds were calm, so still, that the guides said it was the first time they had ever seen reflections in the lakes we passed. Then a rain squall swept in, and suddenly summer disappeared. But soon the sun reappeared and dried out our jackets. One morning, three small birds flitted and flew along beside us as we hiked. Mario said that meant three angels were watching over us.

I felt that my Patagonian odyssey was totally charmed. I knew this the last day as we boarded a bus in El Calafate to return to Chile. It was a blustery, cold, wet, miserable day. It could have been like that during my whole trip, but it wasn't! We rode past a shallow prairie lake full of bedraggled pink flamingoes. We stopped at a roadhouse where Butch Cassidy had hidden out from the law for many months. We arrived back in Puerto Natales in time for me to meet a scheduled English language class, where I enjoyed working with Spanish-speaking high schoolers for several hours. I was back in my teaching element.

The next day, Eduardo delivered me to the airport, as kind and professional as ever. I thanked him profusely as we both wished each other well. I checked in for my flight, but it arrived four hours late, which meant I missed all my connecting flights. But that is a whole other story that I have mostly forgotten in the warm glow of my Patagonia memories.

Patagonia in Argentina | 259

Figure 38. Back packing with the grand children on the Olympic Peninsula

Figure 39. On the summit of Mount Fuji

Figure 40. Mark and me at a memorial on the trek to Everest Base Camp

Figure 41. At a Colorado Trail sign post : Only 100 miles to go?

Figure 42. Expedition to Machu Picchu for a group of one: Me!

Figure 43. Peruvians planting potatoes at 14,000 feet!

Figure 44. Machu Picchu is a breath taking, dramatic site!

Figre 45. Enjoying a boat ride and view after days of trekking beneath these Patagonian peaks

Patagonia in Argentina | 261

Figure 46. Crossing the river on a cable

Figure 47. Ice climbing on a glacier

Figure 48. Fitz Roy massif on the Argentinian trek

Two Months in South Korea

People ask me out of all my travels which place I liked best. I can't answer that question, because no one place is better in every way than every other. Planet Earth is so extravagantly varied and evolving that a location I may have idolized at one time has so changed over the decades that it is hardly recognizable when I see it fifty years later. I too have changed. What I can say, however, is that Korea is the most surprising, fascinating country I have ever lived in, if only for two months.

Part of the reason for this was my almost complete ignorance of the country. Before my stay in Seoul, I can't remember hearing any mention of Korea in school or university. The sum total of my knowledge had been as follows: Korean War, Demilitarized Zone, Seoul, Hyundai, and Japanese invasions. The Koreans whom I met introduced me to a realm of ancient wonders, natural beauty, educational and artistic excellence, and, especially, human kindness. They taught me so much about their country that I felt I had discovered a wonderful secret. My thanks to them all!

In 2005, Hajo accepted a teaching assignment in international relations at Yongsan Army Post in Seoul. I went along as a dependent. We flew out of Denver on October 8. Sometime during the night, we crossed the international date line, landing at Incheon Airport outside Seoul on October 10. Phileas Fogg had gained a day by going from west to east to win his bet of travelling around the world in eighty days. We lost a day, but what we lost in time was compensated for, thanks to the adventures and relationships we made from the first day in Korea to the last.

We immediately settled into our off-base home in Han Shin Building #2, apartment 606. As we arrived, our elderly Korean neighbor greeted us as he looked up from tending his bonsai tree on our shared balcony. The dwarf tree was covered with carefully pruned branches and small green leaves. As our stay progressed, the leaves turned golden and then dropped and blew away. When winter arrived, we would fly home.

In the meantime, we tried to meet the challenges of recycling Korean style. In the apartment-complex courtyard, dozens of bins were clearly marked, in Korean. The Korean groundskeeper watched what was placed in each bin. He saw me coming, kept his eagle eye on me, and swooped over, waving his arms and saying, "No, no, no." Then he showed me exactly where to place coffee grounds and tea leaves, each kind of paper, each type of plastic, each color of glass, each kind of food scraps: some for pigs, some for compost. The Koreans are so dedicated to ecology that they use and reuse metal chopsticks rather than wooden ones in many homes, lunchrooms, and restaurants.

Beyond the apartment courtyard, gingko trees lined the street. They are said to have lived along with the dinosaurs eons ago. Their golden autumn leaves created a canopy of glowing light for the pedestrians below, and there were eleven million in Seoul!

Seoul is a sprawling city, cut by the broad Han River with its twenty-five bridges connecting the two sides of the metropolis. One evening at dusk, I walked down along its banks to find a viewing place for a Korean American fireworks festival. Hundreds of camera tripods staked out spots along the shore. When night fell, thousands of viewers cheered the multicolored explosions and brilliant light showers. It seemed that many experiences around Seoul were crowd-centered extravaganzas!

Even hiking! One day, a group of Koreans and Americans from the Army Post organized a climb up one of the rocky summits in the Bukhan Mountains in the outskirts of Seoul. I was invited to join them. The trail was lined with an ant-like parade of people hiking up rocky trails and scrambling along anchored cables cemented into the bare boulders of the solid bedrock. Hikers, moving up and down, wove under and over and through the arms of others, creating a complicated chain dance to and from the summit. We waited in line for our moment on the tip-top, then scattered out over the rounded rock approaches to find a place to sit and share the view.

Later that day we visited Doseon-sa Buddhist Temple, with its open-air ceiling of fluttering pastel paper lanterns swaying in a light breeze. Hundreds of Korean parents gathered to pray for their children's success in the state examinations for entrance to Korean universities. The belly of a white standing Buddha statue had been rubbed black by the innumerable hands of praying people hoping for good luck for their children. Crowds and more crowds!

Bosung Girls' Middle School

Each morning during our first week in Seoul, I looked out our apartment windows at groups of Korean girls in navy-blue uniforms walking up a long, steep road lugging their books. They were obviously on the way to school. The next Monday, I was ready and fell in step with three of the girls. I introduced myself and asked, "Do you speak English?" They said they studied English, and when I asked, said they would introduce me to their English teacher.

They took me to the teachers' room at the Bosung Girls' Middle School. Their teacher was not there, but the vice principal was. He brought me a cup of tea and made me feel welcome despite our language barrier. Then the principal arrived, and we discovered we could communicate in French. She was very supportive of me during my stay in their school. Soon Miss Sarah Choi, the girls' English teacher, came in. I explained that I was a teacher of English as a foreign language and had taught in France, Germany, Poland, and other countries. I said I would like to volunteer at their school if they could use my services as a native speaker of English. I didn't want to be paid. I just believed in fostering international contacts.

Miss Choi jumped at the opportunity. For the rest of my stay in Korea, I taught at the school three days per week. The first topic was education in American schools. In Bosung School, the girls did much of the janitorial work themselves, scrubbing the floors, washing the windows, dusting the shelves, et cetera. This was to teach them responsibility. We all wore slippers in the school to keep the floors clean. I had never taught in bedroom slippers before, but after the first day it seemed natural.

The Korean girls were a bit envious of American girls because there were so many different clothes and hairstyles and colors in the States. The Korean girls all had straight black hair, wore school uniforms, and had a hard time expressing their individuality. They liked attending an all-girl school because they believed boys would dominate them in a mixed school. They had very little choice of subjects, and were amazed that students in America could choose their classes and participate in after-school activities. The Korean girls usually took extra academic classes after school.

However, the girls did have one after-school activity: producing and presenting the complete musical *The Sound of Music* in English. They were led by Mrs. Koo, another English teacher. The girls played all the parts, male and female, adult and child, with a full orchestra. They presented the show three times, once for Hajo and me along with others who couldn't attend the first two performances. It was a superb production! I wondered if many American middle schools produced such events in a foreign language? Everyone was impressed!

One day, while returning from school to our apartment, I passed a kimchi-pot store. Kimchi is the best-known Korean food: a fermented, salted, seasoned vegetable dish that usually includes cabbage and garlic. Kimchi is stored in brown pottery jars of all sizes. I had previously tried to photograph the pots to show Americans what they were like. I hadn't succeeded in making clear images. On this particular day, there was a whole photography crew taking pictures, so I watched to see what they were doing and how. When they saw me, they immediately stopped their work and came over to me. They were looking for a foreigner to buy a kimchi pot for their coverage as part of the fifty-year anniversary of this kimchi-pot store. Would I please buy a pot so they could photograph me for Korean television?

I agreed, got out my Korean money, and prepared to choose a pot. One of them, who spoke English, told me to put away my money, and thrust American bills into my hand.

The transaction had to be in authentic American money for television. The owner of the store approached me as I explained in English that I wanted to buy a certain kimchi pot. We picked it out, and I paid for it with their American money. She carefully wrapped it in paper and added another small flower vase as a souvenir. I thanked her as I left the store.

The television crew captured it all and were happy with the results. I returned to the store to give back the pot, supposing it was a fake purchase for television. But no. It was mine with their thanks. The next day, my students wondered how I had become a star on Korean television!

Back in the English classes, Miss Choi and the students wanted me to talk with them about race relations in the United States. I asked them what they knew. They had studied part of Martin Luther King Jr.'s "I Have a Dream" speech. It was a good starting point. I gave them a short history of slavery, and then I shared my experiences teaching Native American students in Montana, working with Black and Puerto Rican girls in NYC, and meeting and talking with Martin Luther King Jr. I explained how we were mostly all immigrants in the US.

I asked them if they felt prejudice toward anyone. They said with one voice, "The Japanese!" I asked if they had ever met a Japanese person. None had, and they didn't want to. I told them how a Japanese girl had lived with my family in Colorado for a year. She did much to change people's prejudice against Japan. I hope our discussion at least got them thinking about diversity in a country that is very homogeneous.

The teachers, especially Miss Choi, were very generous and kind. Hajo and I invited her to our apartment for dinner. She arrived with a huge basket of fruit and an invitation for me to spend the next weekend with her. We travelled north to near the DMZ where she had an artist friend, Miss Song, who lived in a restored farmhouse. It had survived the war and bombing and was typical of the old traditional houses, featuring a stucco-walled courtyard with a well and an outdoor cooking area. A roofed path around this patio gave access to each of the rooms. They took me to my bedroom, which was totally bare until they opened two cupboard doors in the wall. They pulled out a futon and feather bed for sleeping on the heated floor. It was delightfully comfortable and cozy even though it snowed during the night.

Miss Song had modernized one room for the inside kitchen and another as an art studio, which had a large worktable made of US Army packing crates. Little was said about the US presence in Korea, but many South Koreans at that time wanted reunification with the North, especially given what they knew about German reunification. After a night in a traditional setting, Miss Choi and Miss Song invited me to their Protestant church for Sunday services. A youth orchestra led the singing, with words projected on the walls, as is the case in many American churches.

About 20 percent of Koreans are Protestant, and 5 percent are Catholic. Others are Buddhists, and many practice a combination of religions. Bosung Girls' School was originally a mission school in the northern part of Korea. It relocated to Seoul after the war to meet the educational needs of girls in South Korea. Shortly after this weekend, I returned to the States and was particularly grateful for my experiences with Bosung Girls' Middle School and the people there.

Jirisan—Sacred Highest Summit

In the meantime, I made several hikes with Joseph, the Korean mountain guide for the US Army post. He invited me to climb Jirisan with him on a Korean-organized event and with a Korean group of hikers. The mountain is located some five hours by bus from Seoul in the south of Korea. This meant getting on a bus at 4:00 a.m. that picked up hikers at several points in the city.

Before the assigned hiking day, Joseph took me by subway to make sure I knew how to get to the pickup point and told me he would meet me there. I paid for my bus ticket and prepared for the expedition, which was a climb of about four thousand vertical feet. The day before our climb, Joseph called me and said he would be unable to go, but that I should go without him. He taught me to say in Korean, "Is this the bus to Jirisan?"

With trepidation, I caught the first morning subway to the bus pickup point. Small groups of hikers waited in the dark. Several buses came by and I asked my one question, "Is this the bus to Jirisan?" The drivers always answered, "No! Wait!" Finally, the correct bus arrived. I climbed on, the only passenger at that stop, but the bus was already very full. I showed the driver my ticket, which was for two people, and in broken English he asked who was the other. I said it was Joseph, who could not come. He did not know a Joseph, and I did not know if Joseph had a Korean name. The driver pointed to an empty seat and as I sat down, the bus sped away.

Across the bus aisle from me, a distinguished-looking Korean man said to me in careful English, "I see your Korean is limited. Perhaps I can help you."

Whew! I was relieved! The man, Mr. Lee, could speak English. I doubt I could have made it through the day without him! He had been in the import–export business until recently, when he lost his job. He was too old to find a new one and too young to retire. He was depressed about his future. His English was quite good. I told him about teaching business English, and he said that he hadn't thought of that. I hope he has since found a new job.

When I met Mr. Lee, he was with a climbing companion. They spoke together a while in Korean, as his friend knew no English. Then he turned to me and said, "Would you permit my friend and me to accompany you to the summit?" How very considerate of them! I was amazed. I said, "Only if I don't hold you back! I would be honored." I relaxed and slept a bit.

Suddenly, the bus slowed and pulled into what looked like a truck stop, but with dozens of buses like ours. The driver parked and called out something that Mr. Lee translated for me: a fifteen-minute rest stop. The bus would not wait. I wondered how hundreds of bus riders could go to the toilet so quickly. I followed the other women hikers into a building. Inside was a huge room with a woman traffic controller in white gloves. On three sides of the room were dozens of toilet stalls with doors. We stood in a line on the fourth side. The minute one woman left a stall, another was directed to the open door. In ten minutes, I was back on the bus.

We travelled on for several more hours as it began to get light. Mr. Lee said we would soon stop at the trailheads. He suggested we climb up the fastest trail on the mountain. There were several different routes that he had climbed, but it was November—the hours of daylight were limited and we were already behind schedule. The climb would take five or six hours. He didn't want to come down in the dark or miss the bus back to Seoul.

When the bus stopped, we literally jumped off and hurried up the road to our trailhead. Mr. Lee said we would hike without stopping for the first hour and then pause for a short water break. I was overdressed but didn't dare stop to take off my jacket, so I kept up the challenging pace. We passed dozens of other hikers on the very rough, rocky trail. When we finally stopped, Mr. Lee said that those we had passed took one look at me and asked, "How old is that woman with you?" They seemed amazed to see a Caucasian woman with white hair climbing a mountain.

I don't remember a lot about the climb, except that the trail was rugged, and the view was grander the higher we hiked. Some hours later, we reached the summit! I felt a real sense of accomplishment, and gratitude for my climbing guides. We sat on large boulders, looking out at the misty vista of the surrounding peaks. It was mysterious and grey with the cold light of late autumn. There were many other hikers spread around us, all talking and snacking, but there was room for all of us. Mr. Lee reached into his pack and said, "Now we celebrate our climb!" He pulled out a large sack of Dunkin' Donuts. I was flabbergasted! The last thing I would have expected, but the Koreans have learned to love American donuts.

Darkness beckoned as we swallowed our last bite of donut with a gulp of water and hurried down the jumbled trail to get back to the bus before nightfall. We used our headlamps to find our bus among the many dozen that had changed position during our climb. I don't think I would have found our bus without Mr. Lee and his friend. When we reached Seoul after midnight, they guided me to a city bus that would take me home since the subway had stopped running for the night. I was overwhelmed by their kindness in going out of their way to help me. I never saw them again, but I have such happy feelings as I remember this day with them, the challenging fun climb, and their generosity.

A Weekend at Jeju Island

Much of the time in Korea, Hajo was busy teaching and I was on my own. But one weekend he was free, and we were able to find a package deal to Jeju Island off the southern coast of Korea. This island is the place of choice for many marriages and honeymoons for Korean couples. It is known for its deep-diving women who regularly collect sea urchins and other mollusks from the ocean depths with only wetsuits and collection baskets. We passed some of these divers on our way from the airport to our hotel. We quickly registered. Then we joined other tourists picking tangerines from the hotel trees . . . a free welcome gift and tasty treat!

The hotel organized tours of various kinds and we signed up for several, but Hajo liked best to relax by a pool and read about Korea. We discovered Jeju had the highest peak in South Korea: in fact, the whole island was this volcanic mountain, Hallasan, slightly higher than Jirisan on the mainland. Hajo said, "Why don't you climb it, Pat?" I didn't need any more encouragement. The next morning, I took a public bus to the trailhead. Hajo's only advice was, "Don't climb it alone." I knew there was no danger of that with the hordes of hikers on every trail I'd seen in Korea.

I started up the broad trail at first light. It was late November, and I remembered from climbing Jirisan that daylight was limited. Just as I thought, there were many hikers. Some hurried past me, and I passed others meandering along. There were big boards indicating exactly how far we had come and how far we had yet to go. Eventually I fell into step behind two couples who were setting a comfortable pace a ways in front of me. I followed them until they stopped at one of the distance boards. When I started to pass them, one man called out, "Hello. You American? My son in Pittsburgh." I stopped and we chatted, him in broken English, until he said, "Please. Come with us."

Byong Rock Choi, who said he spoke "small English," led the five of us up the trail, then up the engineered series of wooden steps and the rugged talus to the summit. We congratulated each other for reaching the top of the old crater, which had the remains of a pond in the bottom below us. I have a photo of me leaning against a fence on the summit with a large sign in Korean beside me, which my companions translated later as "Danger. Do not lean on fence!"

We left the fence and Mr. Choi said, "Now telephone son, Pittsburgh!" A minute later, we were speaking on a cell phone with their son in Pennsylvania! I explained to him in English how much I appreciated his parents welcoming me to join their climb. He had also climbed Hallasan with them. He was a university student studying in Pittsburgh.

I thought we would part ways and descend separately. However, they explained that this was their twenty-fifth wedding anniversary hike for both couples. They had been married on Jeju. They would be honored if I would celebrate with them. We found a grassy area in the sun. They spread a trail map on the grass as a tablecloth, and we shared rice rolls, cookies, tangerines, and other delicacies. It was festive and fun! How kind they were to include me. We descended together to the trailhead. They took a bus one way, and I another.

I soon rejoined Hajo at the hotel. I shared my adventures with him, and he said he had tickets for us to visit Udo, a small rural island off the coast of Jeju, the next day. I especially remember the clear turquoise water washing up on white-sand beaches surrounded by black lava outcroppings: gorgeous contrasts! We enjoyed a local museum, the picturesque farming and fishing villages, and the pristine feeling of the island.

Final Impressions of Korea

All too soon, we were back in Seoul with its eleven million inhabitants and many offerings. Hajo returned to his students, and I to mine. But I had free time to visit the National Museum of Korea. The first room dealt with prehistory in Korea. I was amazed that Korea had more prehistoric stone monuments than any other place in the world. It was known in Asia as "the country of the dolmen." I asked a museum guard about this. She said, "You wait here." A few minutes later, Mr. Kwon, the museum expert on paleontology and the Bronze Age, introduced himself to me in perfect English. We discussed Stone Age structures all over Europe and Asia, and then he gave me a personalized tour of the entire museum for two hours. I later enjoyed visiting a number of the prehistoric table-like structures still standing in and around Seoul.

I frequented the Seoul Arts Center for its many concerts, art exhibits, and classes. Its ballet school had one particularly large practice room, with barres and mirrors and a huge picture window that allowed the public to watch the students practicing as they walked by outside.

My husband and I attended Rotary meetings. We visited the DMZ and learned, among other things, that a railway station had been built on the South Korean side in anticipation of eventual reunification, or at least the free flow of people between the north and south. We visited many of the impressive traditional temples, as well as the numerous shopping centers, in Seoul. The Olympic Park, with its many venues and sculptures, is remarkable.

Unfortunately, our stay was much too short. But before we could leave Korea, we were delayed at the airport. While we were waiting, airline officials invited us into creative art rooms. We were given materials and directions to make a traditional Korean paper box, a fan, and other art objects. The Koreans knew how to make foreigners feel welcome and leave with a good impression of their country.

Kilimanjaro: On the Roof of Africa

In 2012, when I was seventy-four, I got to thinking maybe I should list things I wanted to do before I was too old to attempt them. One image that kept coming to mind was Africa's magnificent stand-alone peak: Kilimanjaro, at 19,341 feet. I had flown past it one night of a full moon on the way to South Africa. It caught my imagination by its sheer size, rising above the African plains where, eons earlier, mankind had come into being. Hajo encouraged me. It would be then, or never.

We started preparations, and he signed me up with Alpine Ascents, a mountaineering company based in Seattle. To help prepare, Hajo and I camped at altitude all over the Colorado Rockies, while I climbed Elbert, Massive, Quandary, and other lesser peaks. I knew that altitude was the main issue for climbers on Kilimanjaro, and all this preparation really paid off. When September came, I was fit and acclimated. Hajo would cheer me on from Colorado.

I flew directly to Arusha, Tanzania, where our climbing group assembled. We visited the outdoor city market, with its local farmers and vendors selling carrots, tomatoes, peppers, oranges, potatoes, and other produce. The wares were spread on cloths lying right on the street. Several sewing machines were set up in the market area on the street for immediate jobs. Shoppers, some in native African prints, others in Islamic clothing, and a few in Western gear, bargained for the best prices. A four-sided white peace column like those seen in many towns around the world expressed in four languages, "May Peace Prevail on Earth." If only it were true.

The next morning, we took a van to the official Machame starting gate. Several groups were assembling their gear and arranging for porters to carry it up the mountain. There are many different routes on such a massive mountain. Some advertised quick climbs, but most people realized that altitude sickness would fell them before they reached the summit. We were encouraged to walk at a leisurely pace ("Pole, pole," our guides chanted—pronounced "Polay, Polay") and spend time at night just acclimating. I was especially happy to see that an American woman was one of our guides. So much of the mountaineering in my earlier life was a man's thing. Still, there were only men working as African guides and porters. The porters' loads were each weighed and shifted around so as not to be over the legal carrying load. I don't know how much that was.

It is fascinating to climb this mountain because you hike through every zone of vegetation. Our drive took us from Arusha through farmlands, pastures, and plantations. Our hiking started in a forest zone with lush vegetation. I saw fog and ferns, and moss growing on trees like in a typical rainforest in the Pacific Northwest. The higher we hiked, the drier it became. We passed through heather and moorland up to the alpine desert above fourteen thousand feet, and finally reached the summit with its arctic conditions and dwindling glaciers. Even in the dry areas with little vegetation, plants and trees unique to the area managed to survive due to their succulent features. I enjoyed seeing vegetation totally unique to Kilimanjaro, and of course, new to me!

I had anticipated having many hours of quiet hiking as we all saved our breath because of the high altitude. Instead, there was a circus-like atmosphere. The porters, more numerous than the hikers, carried ridiculously balanced loads of chairs and tables, of huge cooking pots and pans, of sleeping tents and nondescript bundles of hikers' items. They wove around the hikers on the trail, singing and joking and trying to avoid crashing into each other or the hikers. It was all very fun and funny. There must have been several hundred of us on some sections of the trail.

Eventually, each day, the porters would get ahead of the hikers and set up camp for each organized overnight group in specific destinations. As we paying guests arrived at our campsites, our porters greeted us, singing and dancing and laughing their welcome song. We immediately lost our trail tiredness, found a camp chair, and collapsed in it to watch each member as they were welcomed individually.

On the trail, we mixed with other climbers from all over the world. I enjoyed having a chance to use my French and German with European groups. Several ladies from Germany worked in a retirement home. They wanted to know how old I was and why I was hiking on such a high mountain. They said most Germans after sixty-five retired to a comfortable chair and didn't budge! These ladies were trying to motivate their clients to get up and move! Could they take my picture to show that retired Americans stayed in shape and did challenging things as they got older? They took a photo of me with my backpack and hiking poles, but I told them not to generalize. There were, unfortunately, plenty of lazy, overweight, retired Americans, too.

Kilimanjaro was a kind of microcosm of planet Earth. Overpopulation was the issue. Clean water was so scarce at some camps that the porters had to bring it from several miles away. Most garbage was hauled off the mountain, but human waste was too often left polluting the campsites and boulders beside the trails. Sometimes, the air reeked. That was over ten years ago. Maybe the situation has improved by now.

On the other hand, trekkers were polite and helpful to each other. We all had only what we brought with us. If we needed something, there was always someone willing to oblige. One member of our climbing party had a birthday. Some porters managed to create a birthday cake without an oven and presented it, wearing white baker uniforms they had brought with them from the base for the event. That was one birthday he will not forget!

The highlight for most of us was reaching the summit. The afternoon before our summit attempt, our Alpine Ascents group made camp several hundred vertical feet above the rest of the camps. Long before sunrise, we crawled out of our sleeping bags and tents dressed in every piece of warm clothing we had. It was bitter cold at that altitude. The sky was utterly black, with vibrant stars shivering in deep space. The ground was frozen. The sandy dust around our tents was frozen. There was a feeling of raw exposure.

A quick mug of warm sweet tea improved our outlook. Four of us, with a guide, set off for the summit several thousand feet above us. Our headlamps made dull grey circles on the ill-defined trail as we moved slowly upward. No one spoke. We were each lost in our own thoughts, in our efforts to breathe the thin air, to keep going slowly, slowly. We noticed that the other members of our group were just leaving camp. Hundreds of vertical feet below us, other climbers had begun their trek. They formed small lines of lights, moving in the dark like monks on a silent pilgrimage.

To me, there was something about the whole experience that had a feeling of the sacred. It was like the beginning of time as the sky lit up in the east. As we came over the crest of the immensely broad summit, the sun broke over the horizon far below us, a golden ball of light and warmth. It bathed the plains where humankind first appeared on earth. We stopped and gazed in awe!

We still hadn't reached the summit. Sometime later, the four of us arrived at the summit sign together: "CONGRATULATIONS! YOU ARE NOW AT UHURU PEAK 5895 METERS, TANZANIA." We hugged and congratulated each other. We took photos with the sign. It was a euphoric time. We were the only ones on the summit. Gradually other climbers arrived, and we slipped away to give them time for their moments of glory.

I found a solitary place near the remnants of the glaciers that once formed an ice cap over the whole summit. For an hour, I remained there in the sun, taking it all in. I felt totally happy and thankful for everyone who had been in my life who made this climb possible. It was like they were there celebrating with me: my husband, my dad who had taught me to climb, my mom who had taught me to follow my dreams, my sons and siblings who might never know how I was thinking of them as part of my climb, and the ancestors I would never know but who moved out of Africa to Europe, and generations later to America.

All too soon, I had to join our group and descend to our presummit camp, and then thousands of vertical feet further in one full day to a base camp. An exhausting, long day, but the euphoria of success kept us moving. The following day, from this camp, I joined a safari. So began another adventure in Africa, with many more to follow in unexpected places.

Figure 49. I celebrate on the summit of Kilimanjaro, 19,341 feet

Kilimanjaro: On the Roof of Africa | 281

Figure 50. A typical crowded Korean summit

Figure 51. One of my classes at Bosung Girls' Middle School, Seoul

Figure 52. Kimchi pots in Korea

Eighty Degrees North Latitude

May 28, 2022: the farthest north I shall ever be! After a delay of two years because of the Covid pandemic, I was part of an expedition to Svalbard. Thanks to my son Mark, his girlfriend Joanna, and assorted other folks who invited me to join them, I spent ten days exploring this remarkable destination far off the Norwegian north coast.

Svalbard is a group of islands located where the Greenland Sea meets the Arctic Ocean and the Gulf Stream. We flew three hours straight north from Oslo to the airport near Longyearbyen, the main town, with its two thousand inhabitants from fifty different countries. These hardy souls endure three months of total darkness in winter and three months of continual sunlight in summer. Anyone is welcome to live there, as long as they have a paying job or money to live on, since there is no social welfare system. If they are about to die, they quickly fly south since they cannot be buried in the permafrost. If they are about to be born, their mother flies to Norway to a city with a hospital in case of complications.

The town is powered by coal mined on the islands. The local university students are studying how these islands have literally moved over hundreds of centuries. Continental drift has pushed them northward—from the equator, where the coal formed in the tropics, to their arctic location. Sixty percent of the landmass is covered by glaciers that are quickly melting due to global warming, endangering wildlife in the process.

We joined about one hundred other tourists on the *Ultramarine* expedition ship, which plies the Arctic and Antarctic waters on their alternating summers. That was how we reached exactly 80 degrees, 16 minutes north latitude on the edge of the arctic sea ice. Our ship was not designed to cut through solid sea ice, so our expedition was limited to the western coasts of the islands. These coasts were cut by magnificent fjords and mountains, which provided a voyage of constant sightseeing.

From the ship, we could watch polar bears navigating across snowbanks and glaciers, seeking food for themselves and their cubs. We could wonder at how the walruses crammed together on sandy shores somehow avoided stabbing each other with their sharp tusks. Meanwhile, seals seemed to grin at us from nearby ice flows. During daily shore trips, we enjoyed watching the local reindeer, foxes changing from their white winter coats to their brown summer ones, and thousands of seabirds nesting on island cliffs and the tundra. Sometimes they all took flight simultaneously, a great majestic cloud of beating wings. The warm May sun quickly melted away the winter snowpack, and bright tundra flowers burst into bloom wherever the bare ground appeared. We found scraps of history strewn about, including debris from early mining, exploration, and hunting in the area. We also learned about two failed attempts in the late 1800s to navigate from Svalbard to the North Pole in hot air balloons!

This journey made us ever more aware of the fragile beauty of planet Earth and how global warming is changing our home. It is one thing when you read about it and another when you see it up close, in person.

2023—The Grand Finale

A year after returning from Svalbard I was back in Boulder catching my breath and remembering the remarkable people, places, and experiences that had filled my life. I was looking at my world map, with pushpins marking the localities of hundreds of life-affirming and life-threatening events. I was not ready to call it quits! On my map, I saw a big empty space without any pushpins between Fiji and Africa. Australia and New Zealand were beckoning!

I already knew some things about this land "down under." My neighbor is a native of Melbourne. I have watched the New Year's celebrations from the Sydney Bridge and Opera House on television. I had studied the Australian Aborigines and the New Zealand Māori people. The "K" animals reigned: kangaroos, kiwi, koalas, kookaburras . . . or did they? The Outback, Great Barrier Reef, and Ayers Rock all interested me. But it was all secondhand knowledge. I was seeking the nitty-gritty of discovery in person. My neighbors, who have become great friends, directed me to Overseas Adventure Travel. That sealed it!

I landed in Melbourne. It was early October and spring in the southern hemisphere. The first thing I saw beyond the airport was a Costco store. "I travelled thousands of miles to see this?" Later we flew further to Alice Springs and the Outback. We were bussed to an isolated kind of ranch where we had a choice: sleep in a cabin or out in a swag on the open plateau under the sky. I didn't hesitate. I laid out this traditional Australian large canvas envelope containing my sleeping pad and blankets on a small platform a foot off the ground, where the earth insects couldn't crawl in to share my body warmth.

After a barbecue around a roaring fire, where we ate a tasty kangaroo steak, we retired for the night. I crawled into my swag and lay there looking up at the dark depths of space: 180 degrees of sky from horizon to horizon. No ambient light! A sky so loaded with stars it seemed there was no room for black space between them. I was awestruck. I fell asleep in utter contentment. Sometime in the night I stirred again, wide awake in the palpable awareness of being such a minuscule member of humanity on this vast continent, on our unique planet, moving with our solar system, with our galaxy, through the whole mind-boggling universe spread out above me.

I drifted in and out of sleep. I thought of the Aborigines who had looked into these heavens for the past fifty thousand years. They were true Indigenous people. Why did the European newcomers think their imported culture was so superior? Even during my trip, the non-native majority voted down an Aboriginal request for representation in the Australian government. I felt sad that Indigenous people around the planet have suffered so much over the ages.

The sky lightened in the east. Time to awaken, not to an alarm clock, but to something more elemental: the sun! I slipped out of my swag fully dressed and made my way to the encampment toilet. There in the dusky dawn light, I saw a roll of toilet paper wrapped in its company covering. Name of the company: "Who gives a crap?" Information followed: "Good for your bum! Great for the world! More people have mobile phones than toilets. 50% of profits donated to help build toilets." Thus, I moved from the sublime to the unexpected. Aussie humor!

We visited Ayers Rock (Uluru in the native language), which rose over one thousand feet above us and the surrounding desert. Half of it was reserved for traditional rites and the other half for tourists. We watched our half of the rock transform into a glowing red massif as the sun began to set. Our tour leaders had prepared a table with wine and snacks to celebrate this moving transformation. We raised our glasses.

I was intrigued by Uluru. It has the same geological history as the Boulder Flatirons, which I had climbed some seventy years ago with Pop. Once upon a time, there was a high mountain range. The force of nature eroded it away into sand, pebbles, and larger stones, which washed into the sea. There they were compacted together into layers of cemented sediment. Then a new mountain, building force, created a second range of mountains, which pushed up the sedimentary layers in the process. Under the sea they were horizontal, but they were tipped up to be vertical as erosion ate away the surrounding weaker material and washed it away. In the case of Uluru, this mountain range was far away, but remnants of the strongly cemented sedimentary rock remain in Uluru and other lesser vestiges scattered across the Australian Outback. In Colorado, the Flatirons, Garden of the Gods, Red Rocks, and other sedimentary formations have an identical history, despite being halfway around the world!

Our tour took us to the northeast coast of the continent to Port Douglas. Here we were on the seacoast, looking out toward the Great Barrier Reef. Oak Beach cut a golden scythe of sand along the shore for several miles. I tried to coax a member of our group into walking the pristine stretch with me. They declined. Maybe it was the signs saying to beware of ocean-swimming crocodiles that come on shore to attack unsuspecting tourists! I decided to risk it and enjoyed an idyllic South Sea walk, with no indication of hungry reptiles.

The next day, we took a three-hour cruise out to the Great Barrier Reef. Since Fiji, I was no longer tempted to scuba dive, but we stopped for snorkelling at three different spots. I expected to see vast beds of dying corral. Instead, there were areas of bright-blue staghorn coral and a multitude of other healthy corals. But there was no denying that some coral was a dull gray, and others lay like piles of broken bones. And there were very few fish to be seen. It seemed like a slow death was in process.

We spent several hours at a wildlife sanctuary. We saw koala bears snoozing on tree limbs. I could have had my picture taken holding one for thirty-five dollars. I let him sleep. We saw kangaroos hopping around the grounds . . . the only ones we saw on our entire trip. They were considered a nuisance in Australia, but obviously not where we were! I remembered singing in fifth grade, "Kookaburra sits on the old gum tree / Mighty King of the bushes, he / Laugh Kookaburra, Laugh Kookaburra / Gay your life must be!" I spent half an hour with two of these turquoise-winged fowl, waiting for them to laugh to no avail. But they were handsome guys.

More so than the kiwi bird that we saw in New Zealand. It looked like a rather scrawny brown chicken. The female lays one huge egg, almost half her size, which is incubated and raised by her mate while she recuperates for six months, and then she lays another egg. They are an endangered species because of rats, cats gone feral, and possums that Europeans brought to this part of the world. Previously, the birds had no natural enemies and so lost the ability to fly. There is a lesson in this . . . If you don't use it, you lose it!

I was impressed by Melbourne. It was exceedingly international, and I saw people of all colors and countries wherever I walked. The Royal Botanical Garden took me from desert cacti to New Zealand fern forests, from palm trees to lotus in the wetlands. I met up with my Boulder neighbor from Melbourne, who took me to her favorite pub and along the local beach. A ten-foot-high outdoor mural depicting the smiling face of an Aboriginal child was a "must see" in downtown Melbourne.

In Sydney, there was a double visual extravaganza in the spring. My first view of the Sydney Opera House was through the vibrant purple blooms of jacaranda trees! The stunning reflections of light off the sail-shaped tile structures in the harbor were more captivating than I could have imagined! The interior was one architectural delight after another. Walking through the streets of Sydney, I came upon the city's main cathedral with its doors wide open. I walked into an evening mass sung by a boys' choir: an unexpected inspiration, with the high voices echoing through the Gothic pillars. The officiating priest read a letter from the Primate in Jerusalem praying that peace might soon return to the Middle East. We were one world, both torn in conflict and united in hope for reconciliation.

We arrived in New Zealand in the midst of rugby mania. The All Blacks of NZ were in the world championship against the South African Springboks. Sadly for the Kiwis, South Africa won. Our group guide, who had shared his hope for victory with us, was deflated but continued to lead us through the wonders of his homeland. Green was the color of choice, from verdant pastures to fern forests to jade, a native gemstone. We watched sheepdogs working the flocks across green pastures, and I even got to shear a small part of a sheep. I hiked along numerous trails beneath the spring-green branches. A Māori jade-store owner and I discussed minerals and gems around the world. I bought a simple piece that he had designed. He gave me a natural, smooth, water-washed jade stone he'd found lying on a local beach.

The Māori are the Indigenous people of New Zealand. They populated the two islands starting about six hundred years ago from the Polynesian islands. Their carvings and weaving have developed from their ancestors. While the Aboriginal people of Australia go back tens of thousands of years, with many different cultures and languages, the Māori have one basic language and cultural background.

The relations between the European settlers and the Māori, while still far from perfect, have been easier to foster than those between settlers and the Aborigines. Māori language is taught in many schools, and lands and villages have been returned to Māori ownership and control. Some of the common facial expressions of Māori culture have sometimes been used to intimidate others, for instance, opponents in sporting events. However, a Māori woman explained to me that the bulging eyes also signified "I am seeing what you mean, I am paying attention." The extended tongue meant "Listen to me, I have something important to say."

We visited the city of Christchurch, which was still being rebuilt after two devastating earthquakes. We cruised Milford Sound and saw its majestic mountains and waterfalls and two tiny penguins cavorting on a sandy beach. We left Queenstown after a spring snowstorm covered the mountains in a mantle of white. We concluded our monthlong tour in Auckland, which reminded me of Seattle—built around bays, with skyscrapers and an observation tower sharing the skyline.

Figure 53. Dramatic view of Svalbard

Figure 54. Walruses lounging in Svalbard

Figure 55. Impressed by the Australian outback and Uluru (Ayers Rock)

The Road Not Taken

> I shall be telling this with a sigh
> Somewhere ages and ages hence:
> Two roads diverged in a wood, and I—
> I took the one less traveled by,
> And that has made all the difference.
> Robert Frost

I conclude my journey around the world in eighty years back in Boulder where it began. It has been an amazing adventure! I am filled with gratitude for so many things: For the people of various cultures I've met, whose diverse stories have expanded my perspective, made me laugh, and inspired my faith and actions. For the mind-blowing beauty of nature, from ocean depths to mountain majesty, from intricate seashells to exquisite blossoms. For the discovery of deep space and deep time through dinosaurs and star-studded skies. For the dynamics of music that feed the soul. For surviving the close calls through no strength of my own. And especially for my family and friends who have accompanied me on the expedition of my life so far.

Thanks to you all, and to the reader for joining me on this trip. I've had a grand finale, but I sense there are encores yet to come. And sometimes the encores are the best part!

Acknowledgements

To family, friends and fellow travelers who made these stories possible from our actual life experiences and adventures: There is not room to tell all the remarkable tales we share, or to name each of the ways you have helped create this book. Please know that you are written in my memory, gratitude and affection.

To my writing group associates who urged me over years to write my stories and then to have them published . Without your constant encouragement this book never would have happened.

To publisher Michael Jenet and editor Jessica Medberry and crew who made this published volume a reality: Thank you for your professional expertise and patience in the process of transforming 80 years of scattered stories into this memoir.

About the Author

Patricia (Pat) Peterson is a graduate of Colorado University with honors. She taught in France as a Fulbright Teaching Fellow for two years, been married for 53 years, widowed with 2 grown sons: one a wine maker the other in Public Relations. She taught swimming, French, English as a foreign language, Sociology, elementary, middle school, high school, and university students on various subjects and in various countries: in particular in Colorado, Montana, Haiti, Germany, France, Slovakia, Korea, Japan, and more. She thrives on volunteer work and now, at 86 years old, spends her time on her love of classical music and opera, and believes in keeping active. Her home base remains in Colorado, but she can often be found seeking out new adventures elsewhere.

JOURNEY INSTITUTE PRESS

Journey Institute Press is a non-profit publishing house created by authors to flip the publishing model for new authors. Created with intention and purpose to provide the highest quality publishing resources available to authors whose stories might otherwise not be told.

JI Press focusses on women, BIPOC, and LGBTQ+ authors without regard to the genre of their work.

As a Publishing House, our goal is to create a supportive, nurturing, and encouraging environment that puts the author above the publisher in the publishing model.

Storytellers Publishing is an Imprint of Journey Institute Press, a division of 50 in 52 Journey, Inc.

The world of publishing has changed dramatically. This has also affected authors and their ability to let readers know about their books. Today, most people buy books based on word of mouth.

If you would like to help this author, please consider leaving an honest review of this book on retail sites and book community sites such as Goodreads.